Boneglyph

David C. Sterling

Published by David C. Sterling, 2026.

BONEGLYPH

First edition. July 31, 2026.

ISBN: 979-8233454790

Written by David C. Sterling.

Table of Contents

History is not what happened. History is what remains.

Which means history is always a lie of omission.

— unknown author, unknown age

Absence of evidence is not evidence of absence.

— Carl Sagan

Chapter One

The Great Chronological Lie

On How We Fit Eternity Into a School Textbook

Take a pencil in your hand.

An ordinary one — wooden, yellow, with a pink eraser at the end. About twenty centimeters long. Now imagine that this whole pencil is the history of the Earth. Four and a half billion years, compressed into twenty centimeters of wood and graphite.

The first eighteen and a half centimeters are a dead planet. Lava, cooling seas, poisonous gases, no life at all. Around the sixteenth centimeter, the first bacteria appear. They will reign undisturbed for a long stretch — several centimeters belonging to a world of bacteria alone. Around the seventeenth centimeter, oxygen begins to poison the atmosphere — history's first ecological catastrophe, brought about by life itself.

The dinosaurs appear in the last two centimeters. They vanish in the final centimeter.

And the whole of humanity — from the first chipped stone to your smartphone — is the eraser at the end of the pencil.

Not even the whole eraser. Just its edge. The outer rim that barely brushes your finger.

This is exactly where our problem begins.

✦✦✦

We are creatures with a catastrophically short horizon. Our biological imagination struggles with timespans longer than a few generations. A great-grandfather is already a blur. The Middle Ages are already almost mythology. Ancient Rome is something like a fairy tale with columns and togas. And when the conversation turns to millions of years, the brain simply nods politely and thinks about something else.

This is not stupidity. It is biology. We never needed to know what happened ten million years ago — we needed to remember where the berries grew and where the tiger lived. Evolution equipped us with a tool for survival, not a chronometer of eternity.

But this tool plays a cruel trick on us when we set out to study the history of the planet. We look into the depths of time — and unconsciously overlay it with our own scale. We think of "very long ago" much the way we think of "far away" — as something reachable, fundamentally similar to "here," only farther.

This is the great chronological lie.

Not a conspiracy of scientists. Not a falsification of textbooks. Simply a quiet, imperceptible distortion of scale — as if you were looking at a map of the world where Greenland appears larger than Africa. The Mercator projection does not lie out of malice. It simply cannot fit a sphere onto a flat plane without distortion. In the same way, our minds cannot hold deep time without unwittingly flattening it.

"Geologic time is not simply a very great many years. It is a fundamentally different category of reality, for which human language has no adequate words."

— John McPhee, Basin and Range, 1981

I. The Line That Deceives Us

Open any popular history textbook. Or visit any educational website — it makes no difference. There will almost certainly be a "timeline": a horizontal arrow receding to the left. Along it, neatly placed markers: Egypt, Greece, Rome, the Middle Ages, the Renaissance, the Industrial Revolution, the present day.

It is beautiful. It is easy to understand. And it is an absolute lie about proportion.

Because if you honestly plotted the Cambrian Explosion on that same line — the moment when multicellular life suddenly blossomed into an incredible diversity of forms — it would fall so far past the left edge of the page that you would not find it in the next room. And if you added the appearance of the first cells, you would have to stretch the line clear across the city.

We have drawn history the way it is convenient for us to see it. And we have gradually forgotten that this is a convenience, not a reality.

When a physicist says "long ago," he may mean four billion years. When a historian says "long ago," he means, perhaps, three thousand years. The difference is a factor of a million. Imagine two people who agreed to meet "nearby" — and one stepped out of his apartment, while the other stepped off the planet Earth. They used the same word — but meant utterly different things.

It is precisely this confusion of scales that lies at the root of everything this book has to say.

✦✦✦

Let's try another thought experiment. It's a bit dizzying — I warn you in advance.

Let's take not a pencil this time, but a year. Let's compress the entire history of the Earth into a single calendar year — from January first to December thirty-first.

On January first, the planet forms out of gas and dust. Sometime in February, the crust cools enough for water to appear on it. In March — the first bacteria. April, May, June, July, August — life slowly grows more complex in the ancient seas, almost without visible change. We must wait until September to see the first multicellular organisms.

Fish appear in early November.

The first amphibians come ashore around November twentieth.

The dinosaurs reign from mid-December until December twenty-sixth — and vanish, having received an asteroid under the Christmas tree.

And what about humans?

The entire history of our species — from the first Homo sapiens to the present day — fits into the last thirty-seven minutes of December thirty-first.

Writing appears at about 11:59:30 p.m. Thirty seconds before midnight.

The Industrial Revolution — roughly three seconds before the New Year.

Everything you know from school history textbooks — wars, empires, religions, sciences, arts — is three seconds out of the whole year. Less than a single heartbeat on the scale of the planet.

But here is what's interesting: within this calendar there is an enormous stretch that almost no one thinks about. Between mid-November and mid-December — nearly a month in our conventional units, meaning tens and hundreds of millions of real years — complex terrestrial ecosystems already existed on Earth. Forested swamps. Giant insects. The first reptiles. Entire epochs of biological and ecological complexity.

And not a single hint that anything besides nature left its mark during these epochs.

Or perhaps the hints are there after all — we simply do not know how to read them?

THE SCALE OF DEEP TIME

4,500,000,000 years — age of the Earth

3,800,000,000 years — first traces of life

540,000,000 years — the Cambrian Explosion (dawn of complex forms)

350,000,000 years — height of the Carboniferous (giant forests, giant insects)

66,000,000 years — extinction of the dinosaurs

2,500,000 years — first members of the genus Homo

300,000 years — Homo sapiens

12,000 years — the beginning of agriculture

Ratio of "human history" to the history of the Earth: 0.000067%

II. What "Deep Time" Actually Means

The term "deep time" was coined by the American writer John McPhee in 1981. He was writing a book about geologists and found he had no words to convey what he felt while standing on an outcrop of rock several hundred million years old.

"Deep time" is not simply "a very long time ago." It is a qualitatively different dimension. Physicists, when they speak of quantum phenomena, often warn: do not try to picture it visually, because human intuition is powerless here. Quarks are not "little balls." They are fundamentally different. The same holds true for geologic time — it cannot be "felt" the way people feel the past. It can only be grasped intellectually.

And this understanding changes everything.

When a geologist looks at a cliff face, he does not see merely a handsome rock. He sees a document. Every layer is a page. Every fracture is an event. This person can read time literally with his hands — running a palm over the surface of the stone and understanding that, two hundred and fifty million years ago, this was the floor of a warm, shallow sea. Here — a boundary between worlds, a layer of ash from a volcano that killed off half the species on the planet.

But even geologists are human. Even they are prone to chronological self-deception. Because the brain is built this way: it understands numbers but does not feel them. A hundred million is just a number with eight zeros. Not a reality. Not an experience.

✦ ✦ ✦

Here is a simple test. Try to picture a hundred thousand people. A stadium. This is possible — you have probably been to a stadium, or at least seen one in a photograph. A hundred thousand is real.

Now picture a hundred million people. Already harder. The brain starts to cheat — it simply shows you "a very big stadium." But the true sense of scale is lost.

Now — a billion. The number becomes an abstraction. The feeling disappears.

And now imagine that between you and some event lies not a billion people, but a billion years. Every single year is not merely a unit on a number line. It is winter and summer. It is generations. It is evolution. It is mountains that rose and then became plains again. It is islands that emerged from the ocean and vanished again.

A billion years is roughly forty thousand generations of a civilization like ours. Forty thousand "histories of mankind" — one after another, without pause.

And all of that is only a quarter of the Earth's history.

"Imagine that the whole of human history is a single heartbeat. And the history of life on Earth is a year of continuous beating. Somewhere within that year, perhaps, another heart was beating too."

— Adam Frank, astrophysicist

III. The Blind Spot at the Center of History

There is one period almost never discussed in popular books on the history of life. Its name sounds dull — the Carboniferous Period. Roughly from three hundred and sixty to two hundred and ninety million years ago.

But what was happening on Earth during that time was anything but dull.

Picture a planet covered in gigantic forests. Trees thirty to forty meters tall — and these are not oaks or pines, but something entirely different: giant club mosses, horsetails the size of houses, tree ferns. The atmosphere is saturated with oxygen

— its concentration reaches thirty-five percent, against today's twenty-one. This makes possible what now seems impossible.

In this atmosphere, insects grow to monstrous sizes. Dragonflies with a seventy-centimeter wingspan. Millipedes two meters long. Spiders the size of a cat. This is not fantasy — these are fossils that lie in museum display cases.

In these forests, evolution experimented with forms of life with unprecedented generosity. Here the first reptiles appeared — creatures who, for the first time, broke entirely free of their dependence on water. Here the most intricate ecosystems took shape, a web of interdependencies whose sophistication astonishes even modern biologists.

For more than three billion years, evolution had worked mainly in the sea. Here, in the Carboniferous, it was conquering land at an accelerating pace — and the results of that conquest are staggering.

But no one thinks of this period as a time when something more than mere complex biology could have arisen.

Why not?

Not because it is impossible. But because we have no habit of thinking that way.

✦ ✦ ✦

In 1953, workers at a coal mine in the American state of Illinois discovered something strange. Embedded in a coal seam formed roughly three hundred and twenty million years ago was a metal object of regular shape. It could not have ended up there by chance — the coal had grown around it the way a tree grows around a nail. This meant the object had been there since the time when the future coal was still living wood.

The object was named the "Illinois Artifact" and set aside. No explanation was ever found. The case was an inconvenient one — it fit into no existing category.

This is not the only such case. There are dozens of them — "out-of-place artifacts," or OOPArts, as researchers call them (from Out of Place Artifacts). We will discuss them in detail in the third chapter. But it is important to understand one thing already: not every one of these objects is evidence of an ancient civilization. Nature is quite capable of producing astonishingly "man-made" looking forms.

But to dismiss all of them as mere "tricks of nature" is to fall prey to another bias: the conviction that the impossible is impossible only because it is more convenient for us to believe so.

✦ *A VOICE FROM THE STONE* ✦

I remember how it began.

Not all at once — not in a single day, nor a single generation. First, hands simply learned to stack stones on top of one another. Then they learned to choose — this stone is better than that one, because it is flatter. Then someone realized that a flat stone could be made by hand, if you struck patiently, for a long time, at the edge of another stone.

Millennia passed — or what we took for millennia. Time was measured differently then: by the flooding of the swamps, by the changing color of the sky, by the lives and deaths of those we loved.

We did not know we were living in the Carboniferous Period. We did not even know the word "period." We knew this: here is the swamp, here is the forest, here is a sky so dense and damp that to breathe was to drink.

We did not look like you. But we too looked upward.

And we too asked questions.

IV. Why This Matters, Right Now

Reading this book, you are probably asking yourself: why? Why think about what happened three hundred million years ago? What does it have to do with us, with our lives, with our problems?

A direct one.

We stand at the threshold of an age when, for the first time in human history, we possess the tools to answer the boldest question of all: are we alone in the universe? And not merely in the sense of "is there life on other planets" — but in the sense of "have we ever been alone on this planet"?

This is no idle question. The answer to it changes absolutely everything: our understanding of human uniqueness, our philosophy of progress, our view of the future. If intelligence arose on Earth before — and vanished, leaving no trace — that means intelligence is vulnerable. That technology does not save you. That progress is not linear.

It means that we, perhaps, are not the final chapter of the story. Only the next one.

✦✦✦

There is one more, more practical aspect.

In 2018, two scientists — the climatologist Gavin Schmidt of NASA and the astrophysicist Adam Frank of the University of Rochester — published a paper with an unexpected title: "The Silurian Hypothesis: Would It Be Possible to Detect an Industrial Civilization in the Geological Record?"

They posed a concrete scientific question: if a technological civilization had existed on Earth hundreds of millions of years ago, exactly what might we find today? And what would we never find at all?

Their answers are unsettling. And they are thrilling.

We will discuss the "Silurian Hypothesis" in the fourth chapter — it deserves its own detailed conversation. But it is important to understand, even here, that this paper changed the rules of the game. It turned the question "did an ancient civilization exist on Earth?" from fiction into a legitimate scientific subject.

Because now the question has a methodology. There is a way to search for the answer.

"We are not claiming that an ancient industrial civilization existed. We are claiming that if it had existed, we most likely would not know about it. And that, in itself, is a troubling thought."

— Gavin Schmidt and Adam Frank, 2018

V. Progress as an Illusion of Direction

There is one idea we absorb so naturally from childhood that we never stop to ask where it came from, or whether it is even true.

The idea that history moves forward.

"Forward" in the sense of "toward the better," "toward the more complex," "toward the more intelligent." This is the narrative of progress: from stone to silicon, from cave to skyscraper, from shaman to neurosurgeon. Each successive generation is smarter, more technological, more humane than the last. History is a ladder leading upward.

But where does this idea come from?

From our own experience. Observing a few thousand years of documented human history, we see something resembling progress. And we extrapolate that pattern onto the entire history of life. Life "proceeded" from bacteria to fish, from fish to reptiles, from reptiles to mammals, from mammals to man. A ladder.

But biologists know better: it is not a ladder. It is a bush. A tree with thousands of branches, most of which break off into nothing. Evolution does not "move toward" humanity. It moves toward survival — and that is a fundamentally different direction.

Bacteria have not "gone obsolete." There are more of them on the planet right now than all other living things combined. By the standards of evolutionary success, a bacterium surpasses a human roughly the way a seasoned marathon runner surpasses a child taking his first step.

✦✦✦

This illusion of directed progress is dangerous for one specific reason: it makes us think that intelligence is the inevitable outcome of evolution. That if you give any planet enough time and the right conditions, sooner or later something like us will appear.

But perhaps intelligence is not a law of nature. It is an accident. Rare, fragile, mortal.

Perhaps the Earth has already tried to create intelligence — and failed to preserve it. Or perhaps it succeeded — but we do not know it, because we do not know how to look in the right direction.

Or perhaps — and this is the most unsettling possibility — we do know how to look, but not how to see. Because we are searching for the wrong thing, in the wrong place.

We are searching for something like ourselves.

But what if it was something else entirely?

✦ *A VOICE FROM THE STONE* ✦

Our cities did not look like yours.

We did not cut stone into right angles. We did not build upward — the air here is too dense, too humid, too alive, to reach for the sky. We built outward. We wove ourselves into the forest instead of cutting it down.

Our technology was alive.

Not in the sense that it breathed or grew — though sometimes it did that too. In the sense that it was made from living things. We did not mine metal from mountains — we taught living creatures to do what we needed done. It was slower. It required a patience you do not have, and cannot have, given how short your lives are.

That is why you will not find us.

Not because we never were. But because we became part of what you call nature.

VI. The First Step Toward an Honest View

Recognizing one's own limitations is, perhaps, the most difficult intellectual act a human being is capable of.

Not because it requires special knowledge. But because it requires a particular kind of courage: to admit that the map of the world you have grown used to may be wrong. That some of the "obvious truths" you have never questioned are nothing more than convenient agreements.

One such agreement runs like this: "The meaningful history of the Earth begins with us."

This is not malicious. It is simply anthropocentrism — a view of the world through the lens of humanity. Natural, understandable, inevitable for a species endowed with self-awareness and language. We tell stories — and we place ourselves at the center of our stories. This is normal.

But when we study a planet four and a half billion years old, this reflex becomes an obstacle. It makes us look into the past and see only "scenery set up for our arrival" — instead of seeing self-contained, complete, finished worlds that existed before us and had no need of us for their meaning.

✦✦✦

This book is an attempt to set that reflex aside, if only for a while. To look at the history of the Earth not as a prelude to humanity, but as an enormous, multi-act play, in which our act is the last one — and a very brief one.

We will ask uncomfortable questions. We will examine scientific hypotheses that most scientists do not dare say out loud — not because they have not thought about them, but because academic reputation matters more. We will look at facts that are usually filed away in a drawer marked "anomaly."

And we will be honest. We will not claim to know what we do not know. The difference between "proven" and "possible" is an important one. But "possible" does not mean "impossible." And science — real science — lives precisely in that space between the known and the permissible.

A Carboniferous forest rustles three hundred million years ago. A giant dragonfly cuts through the humid air on wings the size of a dinner table. Somewhere in the canopy of a tree unlike anything you have ever known, something sits and looks down with intelligent eyes.

We do not know what it is.

But we do know we have no right to decide, in advance, that it is nothing.

"Science begins with wonder. But true wonder requires courage — the courage not to know."

— Richard Feynman

VII. The Map We Will Draw

Ten chapters. Ten steps from self-deception to honest astonishment.

In the next chapter we will talk about the planet as a machine of forgetting. About how the "geological eraser" works — the mechanism by which the Earth wipes away every trace of the past with such thoroughness that, a hundred million years from now, nothing of our skyscrapers will remain visible to the naked eye. This is a frightening fact — and, at the same time, the key to understanding why absence of evidence is not evidence of absence.

Then we will examine the "out-of-place artifacts" — strange objects that geology cannot explain. We will be merciless: most of them are observational errors or natural curiosities. But some of them . . .

Then — the "Silurian Hypothesis." That very question two scientists dared to ask out loud: what, exactly, would the trace of a vanished civilization look like? And are we even looking for it in the right place?

Next — a conversation about intelligence. About the fact that intellect is not the monopoly of mammals. About how the creatures that lived on Earth three to four hundred million years ago were biologically far more complex than we have grown accustomed to thinking.

And finally — about ourselves. About what trace we will leave behind. About the "Great Filter" — the riddle of why the universe is silent. And about how that silence may be telling us something very important about the nature of intelligence itself.

✦ ✦ ✦

But first — pick up the pencil once more.

The same one, twenty centimeters long. Set it down on the table. Look at it.

Find the edge of the eraser — right here, at the very tip. Do you see that thin line separating the rubber from the air? That fraction-of-a-millimeter gap is everything you know. Everything written in the textbooks. Every war and peace treaty, every kingdom and republic, every invention and discovery. All the meaning we have managed to invest in the word "history."

Now look at the remaining twenty centimeters of the pencil.

There — silence. An enormous, dark, absolute silence, which we almost never disturb with questions.

It is time we started listening to it.

Chapter Two

The Geological Eraser

How the Planet Destroys Its Own Memory

Imagine a perfect machine for destroying evidence.

It is in no hurry. It harbors no hatred, no intention. It does not know what "evidence" is, or what a "crime" is. It simply works — methodically, blindly, without pause, for four and a half billion years running. It grinds continents to powder, dissolves mountains, buries cities under ash and then uncovers them to the sun again, only to erase with sand and water the little that survived.

This machine is called Earth.

In the last chapter we spoke about how catastrophically small the human perception of time is for taking an honest look at the planet's history. Now it is time to ask the next, far more disquieting question: even if we learned to think on the scale of geologic time — what, exactly, would there be left for us to search for?

The answer is discouraging: almost nothing.

Because the Earth is not a museum. It is not an archive where folders of documents from every era sit neatly filed away. It is a living, seething, restless organism that is constantly digesting itself. And understanding just how merciless that self-digestion is — that is the key to everything the rest of this book has to say.

✦ ✦ ✦

I. A Crust That No Longer Exists

Let us begin with the single most fundamental fact, one rarely spoken aloud outside geology lecture halls: almost the entire original crust of the Earth is gone.

The planet is four and a half billion years old. But the oldest rocks we have managed to find on its surface are fragments of the Acasta Gneiss in Canada, roughly four billion years old, and individual zircon crystals from Jack Hills in Australia, about four and a quarter billion years old. These are exceptions — tiny islands of surviving antiquity in an ocean of vanished matter.

Everything else has been remelted. Not once. Many times over.

The truth is, the Earth's crust is not a permanent shell around the planet but more like a conveyor belt. Tectonic plates are constantly being born at mid-ocean ridges, where molten rock rises from the mantle and hardens, forming new ocean floor. Then, tens or hundreds of millions of years later, those same plates arrive at the continental margins and sink back into the mantle — at subduction zones, where one plate literally dives beneath another and melts in the planet's depths.

This is called the Wilson cycle, after the Canadian geophysicist John Tuzo Wilson, who first fully described this pattern in the 1960s. According to his model, the oceanic crust renews itself completely, on average, every hundred and eighty to two hundred million years.

Read that again.

Every two hundred million years, the entire oceanic crust of the planet — all of it, without exception — sinks into the mantle and is remelted from scratch. Not a single bone, not a single fossilized leaf, not one engraved stone can survive this process if it lies on the ocean floor.

And the oceans, after all, cover seventy-one percent of the planet's surface.

THE SPEED OF FORGETTING

180–200 million years — full renewal cycle of the oceanic crust

71% of Earth's surface covered by oceans

~2–10 cm per year — typical rate of tectonic plate motion

4,000,000,000 years — age of the oldest surviving rocks (under 0.1% of the planet's surface)

0% — share of original oceanic crust older than 200 million years that survives today

Continental crust is somewhat more durable — it is lighter than oceanic crust and so far less willing to sink into the mantle. This is precisely why Earth's oldest rocks are found on the continents, in their geologically stable cores known as cratons. But continents are not immortal either. They crumple into mountain ranges, split apart, drift, collide again, and their surface is ceaselessly worn down by forces we will discuss further on in this chapter.

If, three hundred million years ago, an underwater city had existed on the floor of an ancient ocean — what geologists call Paleo-Tethys or Panthalassa — nothing at all would remain of it today. Not a single brick. The ocean floor on which it stood ceased to exist as a physical object long ago. It melted into the mantle and became new magma, which is perhaps rising right now somewhere beneath a mid-ocean ridge, preparing to become part of a new crust.

The city would not have been destroyed. It would have been remelted into the very substance of the planet.

✦ ✦ ✦

II. Mountains That Become Plains

Fair enough, the attentive reader will say. But not everything lives on the ocean floor. What about dry land? Continents, at least, do not sink into the mantle every two hundred million years — so perhaps something there could survive?

Here begins the second act of the geological eraser's work — and it is no less merciless, though it acts more slowly, more invisibly.

Erosion.

The word is so familiar that we stop feeling its true scale. We think of erosion as something local: a crumbling riverbank, cracks in an old wall, wind whittling away at a desert rock. But on a scale of tens of millions of years, erosion is a force capable of wiping entire mountain systems off the face of the planet.

Take the Appalachians — a mountain chain in eastern North America. Today they are low, forested hills, popular with hikers and utterly unintimidating in height. But geologists know that once, around three hundred million years ago, the Appalachians rivaled the modern Himalayas in height. Perhaps they were even taller.

What happened to them? Nothing dramatic. Just rain. Just wind. Just ice, which every winter crept into cracks in the rock and split them apart from the inside as it froze and expanded. Just rivers, which year after year, millennium after millennium, carried grains of broken stone down toward the sea. Just time.

Three hundred million years — and all that remains of a mountain system as tall as the Himalayas are green hills where backpackers now stroll, never suspecting they are walking on the ruins of a mountain giant.

If erosion can grind a Himalaya-scale range down to gentle hills, what would it do to any man-made structure, given enough time?

"Give me enough time, and I will turn a mountain to sand with nothing but a single drop of rain."

— a geological principle, paraphrased, traceable back to James Hutton

The average rate of erosion on continents runs from several tens to several hundred meters per million years — depending on climate, relief, and rock type. That sounds slow. But over a hundred million years — precisely the timescale that concerns us if we want to know whether a prehistoric civilization could have existed on Earth — this "slow" force strips several kilometers off the surface of a continent.

Several kilometers.

Consider what this means. Any structure, any foundation, any underground tunnel built on today's land surface will be entirely stripped away by erosion — like the peel off an apple — within a mere few tens of millions of years, unless it happens to be buried under new sediment before erosion reaches it.

And if it does get buried? Then a third force enters the game.

✦ ✦ ✦

III. Burial and Metamorphism: The Second Act of Destruction

One might suppose that burial also has a protective side. After all, it is precisely because of burial under layers of sedimentary rock that dinosaur fossils have survived — bones seventy, a hundred, a hundred and fifty million years old, which we can view in museums today.

That is true. Burial is the only real chance of preserving anything across geological timescales. But that chance comes with brutal conditions attached.

First, the statistics of fossilization are merciless. Paleontologists estimate that fewer than one ten-thousandth of one percent of all living organisms that have ever existed on the planet become fossils. For a bone to turn into a fossil, a rare confluence of circumstances is required: the organism must die in an environment where sediment quickly buries it — silt on a lake bottom, sand in a river delta, ash from a volcanic eruption. It must not be eaten by scavengers. Minerals carried in groundwater must replace the organic tissue before it decays. And then — hardest of all — that patch of crust must not later be eroded, melted, or deformed enough to destroy the fossil.

Each of these conditions is a filter. And filters, applied one after another, yield a catastrophically small probability.

This is exactly why fossils are not a "random sample" of ancient life. They are a tiny, skewed, incomplete selection of the few organisms lucky enough to die in the right place at the right time. Entire groups of creatures, entire ecosystems, may have left nothing behind at all — simply because they lived under conditions unfavorable to fossilization.

But even if an organism — or an artificial object — is successfully buried and begins turning into a fossil, its ordeal is only beginning.

Sedimentary rock keeps accumulating on top. With every new layer, pressure and temperature rise. Several kilometers down, the rock begins to change its structure — minerals recrystallize, the original texture is erased. This process is called metamorphism, and it can destroy even what has survived millions of years of burial in relatively undisturbed condition.

Limestone becomes marble. Sandstone becomes quartzite. Shale becomes gneiss. And in the course of this transformation, any fine structure — any detail, any shape, any hint that what lies before us was something artificial rather than natural — dissolves into the rock's new crystal lattice.

Metamorphism is forgetting at the molecular level.

✦✦✦

IV. Continental Collision: A Press of Planetary Scale

There is one more process worth discussing on its own — because it explains why even buried, surviving structures are not guaranteed to escape destruction.

Continents do not stand still. They drift — slowly, but inexorably, obeying the same tectonic forces that grind up the ocean floor. And from time to time, they collide with one another.

A continental collision is one of the most colossal geological events imaginable. When India broke away from the ancient supercontinent Gondwana and, after tens of millions of years of wandering, crashed into Asia, the collision gave birth to the

Himalayas — and continues to lift them still, at a rate of about five millimeters a year.

But what happens to the rocks caught in the path of that collision? They are crumpled, broken, turned inside out, thrust up several kilometers high — and then immediately set upon by the same erosion we discussed above. Layers that once lay flat on the seafloor now stand vertical, or upside down. Rocks that formed in one place end up shoved hundreds of kilometers from where they started.

The Earth's history records at least several episodes of supercontinent formation — Rodinia, Pannotia, Pangaea — each time resulting from the collision of nearly all of the planet's continental mass into one gigantic landmass, followed by a fresh breakup and scattering. Pangaea, the most recent such supercontinent, began forming right at the end of the Carboniferous Period, around three hundred million years ago — the very period we discussed in the first chapter.

This means that any civilization of the Carboniferous would have witnessed not merely the slow erosion of its world, but a colossal tectonic cataclysm crushing continents together into a single supercontinent.

It is hard to imagine a more effective way to erase the material traces of a civilization than to ram the continent it lived on into another continent, slowly raise the resulting mountains several kilometers high, and then subject them to erosion for the next hundred million years.

✦ *A VOICE FROM THE STONE* ✦

The earth trembled for weeks before it happened.

The old ones said it had happened before — many generations back, before their grandfathers had even learned to count generations. The earth trembles, then it cracks, then where there was water a wall rises, and where there was solid ground a chasm opens.

We did not build in stone the things we feared to lose. We learned this early — earlier than we learned to count years. Everything that mattered, we carried inside ourselves: in memory, in songs, in patterns passed from generation to generation by hand, not by chisel.

Perhaps that is why you will not find us.

We left no monuments. We knew something you do not yet know: stone is a poor keeper of memory. It remembers only for a short while. Only what lives, and passes itself on to those still living, remembers for long.

V. Volcanism: The Seal of Remelting

To the list of destroyers we must add one more participant — perhaps the most spectacular of them all.

Volcanic activity has not been uniform throughout Earth's history. There have been epochs when volcanism reached catastrophic proportions, almost unthinkable by modern standards. The most famous example is the Siberian Traps, a gigantic outpouring of basaltic lava that occurred roughly two hundred and fifty-two million years ago, at the boundary between the Permian and Triassic periods.

This event covered an area comparable in size to Western Europe with lava. The volume of erupted material is estimated at several million cubic kilometers. The eruption continued, with interruptions, for roughly a million years and, in the view of most researchers today, was the principal cause of the greatest mass extinction in the planet's history — the "Great Permian Extinction," which wiped out, by various estimates, between ninety and ninety-six percent of marine species and roughly seventy percent of land species.

If cities, roads, or the infrastructure of some civilization had once existed on the land now covered by the Siberian Traps, they would have ended up buried beneath kilometers of hardened lava. And that would be the best-case outcome: if the lava had poured out directly where they stood, everything that could burn would have burned. Everything that could melt would have melted.

Volcanic ash and lava are not merely destroyers. Sometimes they become perfect preservers — think of Pompeii. But far more often, they become perfect annihilators, leaving behind only a uniform layer of hardened rock that holds no trace of whatever came before.

THE GREAT EXTINCTIONS AND GEOLOGICAL CATACLYSM

~444 million years ago — the Ordovician extinction

~372 million years ago — the Devonian extinction

~252 million years ago — the Permian extinction ("the Great Dying," up to 96% of marine species)

~201 million years ago — the Triassic extinction

~66 million years ago — the Cretaceous–Paleogene extinction (death of the dinosaurs)

Each of these events was accompanied by large-scale geological restructuring of the planet's surface

✦✦✦

VI. Glaciers: Bulldozers of the Ice Age

There is one more tool in the arsenal of the geological eraser that deserves separate attention, because it works differently from everything we have discussed so far. It does not melt. It does not crush mountains with the force of continents. It scrapes.

Ice ages are no rarity in Earth's history. The planet has endured them many times over, including, according to some geologists, episodes of nearly complete surface glaciation — a hypothesis known as "Snowball Earth," dating to the period between roughly seven hundred and six hundred thirty-five million years ago.

Glaciers move slowly — meters, tens of meters a year. But ice kilometers thick carries such mass and pressure that it acts as a giant abrasive tool, shearing off the upper layers of rock in its path, grinding stone into sand and clay, and carrying debris hundreds of kilometers from where it originated.

When the North American ice sheet retreated after the last glacial period, roughly ten to twelve thousand years ago, it left behind a landscape literally scraped down to bedrock in some regions — the Canadian Shield, for instance. Everything that had once existed on that surface was either carried off by the ice or ground into sediment and scattered across a vast area.

A single ice age is enough to wipe away everything that accumulated on the surface over tens of thousands of years. And in the planet's history, there have been dozens of such episodes, counting all known glaciations since the Archean eon.

✦✦✦

VII. The Arithmetic of Forgetting

Now let us gather all these forces together and see what happens when we apply them, one after another, to a hypothetical civilization that existed on Earth, say, three hundred million years ago.

If its structures stood on the ocean floor — they are remelted into the mantle within the following two hundred million years. A hundred percent certain, no exceptions.

If they stood on land, were never buried, and lay in the path of a tectonic continental collision — they are crumpled, thrust up into mountains, and worn away by erosion within the following one to two hundred million years.

If they stood on land and were buried under sedimentary rock — they come under mounting pressure and heat, and several kilometers down they begin to metamorphose, losing their structure, within the following tens of millions of years — unless tectonic forces lift them back to the surface, where erosion takes over.

If they lay in the path of a large-scale volcanic event — they are melted or buried under lava and ash.

If they survived all of the above on some especially stable patch of continental crust — an ancient craton — they are still subject to slow but relentless erosion, at a rate of several kilometers per hundred million years, unless they lie deep enough beneath the surface.

Multiply all of this by three hundred million years, and the probability that anything recognizable would still survive on the planet's surface today approaches zero.

"The absence of fossil evidence for an industrial civilization does not rule out its existence a hundred, two hundred, or three hundred million years ago. The geological record is simply not complete enough to give a definitive answer."

— Gavin Schmidt and Adam Frank, "The Silurian Hypothesis," 2018

This conclusion is not the speculation of conspiracy theorists. It is a finding published in a peer-reviewed scientific journal, reached by two respected scientists, one of whom, at the time of publication, headed NASA's Goddard Institute for Space Studies.

And this conclusion overturns the very logic of the search. We are used to thinking: if an ancient civilization had existed, we would already have found its traces. But geology tells us the opposite: if a technological civilization had existed a hundred or more million years ago, we would most likely have found almost nothing. Not because it was impossible. But because the planet operates like a machine specifically — though with no intention at all — built to erase its own history.

✦✦✦

VIII. What Survives — and Why It Matters

Does all of this mean that searching is pointless? That any evidence has certainly been destroyed and the question is closed?

No. And here is why.

The geological eraser works selectively. It is devastatingly effective against physical objects — buildings, tools, skeletons. But there are categories of traces it erases far more slowly, and sometimes barely touches at all.

Isotope ratios of elements dissolved in ancient rocks. Chemical anomalies recorded simultaneously in sedimentary layers around the world. Sharp, unexplained spikes in carbon-13 concentration or rare metals in the geological record of a particular epoch. Traces like these do not require a building to survive. They require only that, somewhere on the planet, in a sufficiently protected spot — the floor of a deep-sea basin, within a body of sedimentary rock that escaped metamorphism — a thin layer survived, recording the chemical "exhaust" of a civilization, even if the civilization itself vanished without a trace.

This is exactly what the "Silurian Hypothesis" is built on, which we will examine in detail in the fourth chapter. If the geological eraser destroys objects but does not always erase chemistry, then the right question is not "where should we look for buildings?" but "where should we look for isotopes?"

This changes the entire direction of the search. And it is the only sensible response to the merciless arithmetic of forgetting we have just laid out.

✦ *A VOICE FROM THE STONE* ✦

We knew it was temporary.

Not because we were wiser than you. Simply because the world we lived in was more honest than yours — it never let us forget, not for a single day. The earth cracked before our eyes. Seas came and went within a single generation's lifetime — or so it seemed to us, though it was probably long centuries; we simply did not count time the way you do.

We did not try to remain forever. We tried to live well the span we were given.

Perhaps that is the difference between us and you. You build as though you believe the world will stand forever. We knew, with certainty, that it would not.

And we were right.

IX. A Lesson in Humility — and in Hope

There is something deeply sobering in realizing just how thoroughly the planet destroys its own memory. But within that sobering thought lies a peculiar, strange hope as well.

If the geological eraser is this merciless toward any material trace, then our own lack of evidence for ancient civilizations proves nothing. We cannot say: "We looked and found nothing, therefore they never existed." We can only say: "We looked for the wrong thing, in the wrong way, if what we hoped to find was something capable of surviving hundreds of millions of years."

This does not mean that an ancient civilization existed. Caution here is not merely good manners but a scientific necessity. But it does mean that the door many consider closed on the grounds of "absence of evidence" actually remains ajar — because the very nature of the evidence we might hope to find turns out to be nothing like what our intuition assumes.

In the next chapter, we will finally turn to specific cases — to those very "out-of-place artifacts," strange finds that surface now and then from the depths of the geological record and leave scientists scratching their heads. We will examine them honestly: without sensationalism, but also without hasty denial. Because now, armed with an understanding of how the geological eraser works, we can ask a far more precise question.

Not "why haven't we found any cities?" — but "what could possibly have survived at all, and would we even recognize it if we saw it?"

"We are not searching for absent evidence. We are searching for evidence of absence — and that is a war science can almost never win outright."

— a paraphrased principle of scientific skepticism

The planet keeps no museums. It preserves only what happens to survive in the cracks of its ceaseless self-renewal. And the task of the next chapter is to look precisely into those cracks.

* * *

To be continued in Chapter Three: "Artifacts Out of Time (OOPArts)"

Chapter Three

Artifacts Out of Time

A Catalogue of Things That Were Not Supposed to Exist

In 1968, a California collector named Wallace Lane sawed in half a stone he had found in the desert.

He expected to find a cluster of crystals inside — as is usually the case with the rounded concretions locals call "geodes" and happily sell to tourists as souvenirs. Instead, the saw struck something hard and unmistakably man-made: a ceramic cylinder with a metal rod inside.

Further examination showed it was a spark plug, manufactured, judging by its design, sometime in the 1920s. An ordinary part of an internal combustion engine.

There was just one problem: the rock this part had seemingly "grown into" looked, on the surface, like a fossil hundreds of thousands of years old.

Thus was born the legend of the "Coso Artifact" — one of the most famous "out-of-place artifacts" of the twentieth century. We will return to it in detail later in this chapter, because its solution turns out to be every bit as instructive as the find itself was sensational.

But first — let's define our terms.

✦✦✦

I. What OOPArt Means, and Why It's an Awkward Word

The term OOPArt — short for Out of Place Artifact — was coined in the 1970s by the American cryptozoologist Ivan T. Sanderson. He introduced it to describe finds that, in his view, contradicted the accepted chronology: objects that seem too modern, too sophisticated, or too technological for the geological era in which they were supposedly found.

The word took hold — and confusion took hold right along with it. Because today, everything from fakes and hoaxes to genuine, well-documented scientific puzzles that simply have no convincing explanation yet gets lumped together under the OOPArts umbrella.

This chapter is an attempt to bring some order to that chaos. We will examine several of the most famous cases that usually come up in conversations about "ancient technology," and apply the same rigorous question to each one: what do we know for certain, what is likely, and what is pure speculation?

This is uncomfortable work. It is far more pleasant either to believe everything with wide-eyed enthusiasm, or to sneer condescendingly at all of it. But genuine curiosity demands the worst of both worlds: a patient examination in which some cases turn out to have dull, unambiguous answers, while others remain genuinely, unsettlingly unclear.

"Extraordinary claims require extraordinary evidence. But that does not mean extraordinary questions do not deserve an answer."

— a principle paraphrased from Carl Sagan

II. The London Hammer: A Stone That Holds a Tool

In 1936, a married couple named Hahn were walking along a creek near the town of London, Texas. Among the rocks, they noticed something protruding from the stone — a wooden handle. Inside the rock, encased in a limy crust, was a metal hammer.

The find quickly grew into a legend. The hammer was declared "proof" that humans — or something like humans — existed on Earth long before the accepted dates. Some enthusiasts dated the rock the hammer was "embedded" in to the early Cretaceous — a rock more than a hundred million years old.

The reality turned out to be far duller — and in that dullness lies the single most important lesson of this chapter.

The London Hammer, as detailed geological and metallurgical analysis has shown, is a perfectly ordinary nineteenth-century tool, with a wooden handle and an iron head made using methods typical of that era. The "rock" surrounding it is not ancient bedrock at all, but a concretion — a mineral formation that can build up around any object when water rich in calcium carbonate deposits minerals rapidly.

This is neither rare nor mysterious. Museums around the world hold concretions that formed around tin cans, bottles, and even human bones — and forming such "capsules" can take not millions of years but literally a few decades, given the right conditions: a high concentration of dissolved minerals and a stable, damp environment.

The age of a concretion is not the same as the age of the rock it is found in — and it is certainly not the same as the age of the object it formed around.

This is one of the most common errors in conversations about OOPArts: confusing the time it took the shell to form with the age of what's inside it. The London Hammer ended up in mineral-rich water and was, fairly quickly — almost instantaneously by geological standards — "packaged" inside a concretion. There is no paradox here. There is only a piece of evidence misread.

✦ ✦ ✦

III. The Klerksdorp Spheres: A Perfect Shape Without a Maker

Far more interesting is the case known as the "Klerksdorp spheres" — small, nearly perfectly round metallic objects found by South African miners in the rocks of the Ongeluk Formation near the town of Klerksdorp. Geologists estimate the age of these rocks at roughly three billion years — making them among the oldest rocks on the planet available for study.

The spheres, ranging from a few millimeters to a few centimeters in diameter, some with distinctive parallel grooves around their equator, became a genuine sensation among proponents of ancient technology. Their perfect roundness and unusual hardness — reports claimed the material was so tough it was hard to break even with modern tools — seemed incompatible with any notion of a random natural origin.

However, the geologists who studied these finds offered an explanation that, while less sensational, is remarkable in its own right: the spheres are concretions of pyrite and hematite — minerals that, under certain crystallization conditions, genuinely can form nearly perfect spherical shapes. This phenomenon is well documented at other deposits around the world and is known as "pyrite spherules" or hematite concretions.

As for the "grooves" — they are a natural consequence of the mineral's crystal structure, not marks of mechanical tooling. Nature, once again, turns out to be capable of a geometric precision that strikes the human eye, intuitively, as man-made — even though it is not.

This does not mean the Klerksdorp spheres stop being remarkable. Three billion years is a staggering age — nearly two-thirds of the planet's entire history. But the wonder here belongs not to some hypothetical civilization, but to mineralogy — the science of how inanimate matter can produce forms our brains instinctively read as the product of design.

A PRINCIPLE WORTH REMEMBERING

Geometric regularity, by itself, is not proof of intelligent origin.

Crystals, concretions, basalt columns (the Giant's Causeway in Ireland, for instance) and countless other natural structures display near-perfect symmetry without any involvement of a mind at all.

The only reliable test of artificiality is not shape, but internal organization that natural processes cannot explain: metal alloys not found in nature; traces of directed, non-random tooling at the microscopic level; repeating, standardized elements indicating serial manufacture.

IV. The Wolfsegg Iron Cube: A Meteorite Wearing a Mystery's Mask

In 1885, at an Austrian foundry in Wolfsegg, workers splitting a block of lignite roughly twenty million years old discovered a strange object: a nearly perfect iron parallelepiped, with rounded corners and a distinctive groove down the middle, resembling a mold seam.

The find became known as the "Wolfsegg Cube" (sometimes the "Salzburg Cube," after the place where it was later exhibited) and remained for decades one of the most debated puzzles in alternative-history circles. Twenty million years — long before humans existed — and yet here was an object strangely resembling an industrial product.

Modern chemical and structural analysis has produced a far more prosaic, but no less interesting, answer: the cube's composition — iron with a high nickel content, characteristic of meteoritic material — is nearly identical to the chemical makeup of many iron meteorites. Its near-parallelepiped shape is also well explained by a meteorite's natural tendency to split along the planes of its crystal structure — a phenomenon known in so-called octahedrites, meteorites with a distinctive Widmanstätten pattern.

The "seam" that so resembled a mold line appears, in all likelihood, to be a natural fracture or the result of oxidation along a crystallographic plane.

As for the coal seam in which the cube was found, two fairly mundane explanations are possible. Either the meteorite fell to Earth long before the coal deposit formed and was later covered over by plant matter that turned into coal; or — more likely, given the circumstances of the nineteenth century, when scientific documentation of finds was rarely rigorous — the object ended up in the coal seam much later, as a result of ordinary human activity, and its geological context was established mistakenly or after the fact.

The Wolfsegg Cube is a fine example of how the absence of modern dating and analytical methods at the time of a discovery can breed a legend that rigorous re-examination later largely dissolves. Though, in fairness, it should be noted: the original specimen itself was lost sometime in the twentieth century, and some of the modern research relies on replicas and archival descriptions — meaning full scientific clarity has never quite been achieved here.

✦ ✦ ✦

V. The Coso Artifact: Anatomy of a Modern Hoax

Let's return to the story that opened this chapter.

After Wallace Lane and his colleagues sawed open the stone found in the Coso desert, California, and discovered the ceramic-and-metal object inside, a sensational version of events leaked into the press and scientific circles: a "fossilized" mechanical part as much as five hundred thousand years old.

This claim failed to withstand scrutiny almost immediately. X-ray analysis conducted shortly after the story broke, and later, more detailed studies, showed the object to be a Champion-brand automobile spark plug manufactured in the 1920s — complete with its ceramic insulator intact and the remains of a corroded metal housing.

The so-called "fossilized rock" around it turned out not to be a bedrock geological formation at all, but a dense conglomerate of clay, pebbles, and metal corrosion products — in other words, a formation that built up around the part over the mere decades it spent in mineral-rich soil, under desert conditions that favor the rapid formation of hard mineral crusts.

The story of the Coso Artifact has become something of a textbook case in OOPArt literature — not because it confirms the existence of ancient technology, but because it flawlessly illustrates the central methodological hazard of this entire field: the desire to see a mystery where plain scientific skepticism and elementary fact-checking would suffice.

Had Lane and his colleagues simply consulted independent specialists in early-twentieth-century automobile parts instead of announcing a "prehistoric mechanism," the answer would have been found within a single day, not decades of heated debate.

"Before reaching for an extraterrestrial or prehistoric explanation, one should always exhaust every earthly and modern explanation first. That is what's called the scientific method — not pedantry."

— a widely held principle of critical thinking in paleontology

VI. The Aiud Aluminum Wedge: An Unsolved Case

Not every case ends in so tidy a resolution. And honesty requires admitting this openly.

In 1974, on the bank of the Mureș River near the Romanian town of Aiud, excavator operators working at a construction site discovered a metal wedge-shaped object amid a layer of sand that also contained the bones of a mastodon — an extinct relative of elephants.

The object was handed over to the Institute of Archaeology in Cluj-Napoca, where it underwent chemical analysis. The result was strange: the wedge was made of an aluminum alloy containing traces of copper, silicon, zinc, lead, tin, zirconium, cadmium, nickel, bismuth, and gallium — a combination that, according to some researchers, is highly uncharacteristic of any accidental natural aluminum alloy and points instead to an industrial origin.

The subsequent fate of this case grew tangled. The object itself is partly lost: one piece is held at the Transylvanian History Museum, but the context of the find — the precise layer in which the wedge was discovered — is documented extremely poorly, since the discovery was made by construction workers on the job, not during a controlled archaeological excavation.

The dating of the surrounding sediment, in which both the wedge and the mastodon bones are said to have been found, is estimated by some geologists at roughly twenty thousand years — which is not geologically deep antiquity at all (on the scale of this book, it is a literal instant), but still long predates any known aluminum industry, since the electrolytic production of aluminum was not invented until the late nineteenth century.

Skeptics point to an obvious alternative explanation: the part could have belonged to modern equipment — the leg of an excavator, say, or a drilling rig — and ended up in the sediment by accident, during the very construction work in question, well before it was "discovered" in a context later linked, after the fact, to the mastodon bones.

The trouble is, neither version — the sensational one or the skeptical one — is backed by documentation rigorous enough to satisfy strict scientific protocol. The Aiud Wedge remains, perhaps, the most frequently cited yet least convincingly resolved case in the entire history of OOPArts, precisely because confirming and disproving its origin today are equally difficult.

✦ *A VOICE FROM THE STONE* ✦

We made things that did not look like stones.

Not much — only what was necessary. We did not cover the earth in metal, as you, it seems, are about to do. We took from the earth exactly as much as we needed to survive until the next rainy season, to raise our children, to bury our old ones with dignity.

What is left of it? Not much, probably. Whatever little remains likely lies now deep beneath what you call the crust — or was melted down again, becoming part of a new mountain, a new sea, a new world that has no idea what it is made of.

Sometimes I want to tell you: do not look for our things. Look for our ash. It is dissolved in your own air, in your own bones. We did not vanish without a trace — we became part of what became you.

VII. The Illinois Pot: Back to the Coal

In the first chapter, we briefly mentioned the "Illinois Artifact" — a metal object reportedly found in 1953 in a coal seam roughly three hundred and twenty million years old. It's time to examine it more closely, because this case beautifully illustrates the full difficulty of working with finds of this kind.

The problem starts with the source itself. Unlike the London Hammer or the Coso Artifact, for which museum specimens exist that are available for independent re-analysis, most reports of "artifacts in coal" from the nineteenth and early twentieth centuries have reached us solely as newspaper items and secondhand retellings — often with no surviving physical specimen, no precise find location, and no geologist's name to confirm the context.

This does not automatically mean all such stories are fabrications. But it does mean that, from a scientific standpoint, most of them cannot be verified — and an unverifiable claim, however intriguing, cannot be treated as evidence.

Even so, some such cases have a fully plausible geological explanation that requires neither hoax nor ancient civilization. Coal forms from plant remains in swampy environments — and swamps frequently swallow all sorts of objects, which can slip into cracks in the rock long before miners ever expose those cracks. What's more, concretions of pyrite — "fool's gold" — can form dense, metallic-looking formations of odd shapes within coal seams, which a non-specialist could easily mistake for worked metal.

The absence of a surviving specimen is not proof that the find never happened. But neither is it proof that it was exactly what the legends describe.

✦✦✦

VIII. How to Tell a Wonder From an Illusion: Five Questions

By this point, the reader probably feels a legitimate irritation: almost every case we've examined ends in disappointment. The hammer — a nineteenth-century tool in a concretion. The spheres — mineral concretions. The cube — probably a meteorite. The spark plug — just a spark plug. Only the Aiud Wedge remains genuinely unresolved.

This is neither an accident nor a result of the author's bias. It is statistics. The overwhelming majority of "sensational" finds, when properly checked, turn out to be explained by natural processes, dating errors, or plain human factors — confusion, a hunger for sensation, and sometimes outright deception for fame or money.

But this is exactly why it's important to lay down clear criteria — a kind of filter that separates a genuinely significant anomaly from yet another concretion, plus an imaginative journalist thrown in for good measure.

Here are five questions worth asking of any such find:

First: has a physical specimen survived that is available for independent, repeatable analysis using modern methods — rather than just an archival description or photograph?

Second: was the geological context of the find — the precise layer, depth, and composition of the surrounding rock — documented by a specialist on site, rather than reconstructed from memory or conjecture after the find became a sensation?

Third: have the simple alternative explanations been ruled out — a concretion, a meteorite, accidental modern contamination, a dating error?

Fourth: does the object have a structure that cannot, in principle, be explained by any known natural process — a metal alloy not found in nature, say, or traces of directed, non-random tooling at the microscopic level?

Fifth: is the find reproducible — that is, do similar, independently documented cases exist elsewhere, or is this a unique, one-off specimen with nothing to compare it against?

Applying this filter to the cases examined in this chapter, we find that almost none of them satisfy all five criteria at once. And this — not any "conspiracy of official science" — is the principal reason serious scientists treat the OOPArts catalogue with restrained skepticism.

FIVE QUESTIONS FOR ANY ANOMALOUS FIND

1. Has a specimen survived for independent verification?

2. Was the geological context documented on site?

3. Have the simple alternative explanations been ruled out?

4. Is there structure unexplainable by natural processes?

5. Is the find reproduced in independent, analogous cases?

✦✦✦

IX. Why This Chapter Does Not End in Disappointment

It might seem as though this chapter's conclusion is the typically skeptical one: "everything's explained, nothing to see here, move along." But that would be a misreading.

The real lesson of this chapter is not that mysteries don't exist, but that our tool for recognizing genuine mysteries is remarkably crude and easily fooled. The human brain evolved to see design where there may be none — psychologists call this "pareidolic apophenia," the tendency to find meaningful patterns in randomness. It's the same mechanism that makes us see faces in clouds or in electrical outlets.

This very tendency has led several generations of researchers to declare a sensation what turned out, on closer inspection, to be a concretion, a meteorite, or a car part. But the reverse is also true: that same caution, taken to an extreme, risks making us wave away the rare cases that, like the Aiud Wedge, genuinely resist easy explanation — simply because "surely this is some kind of mistake too."

The right stance lies neither in belief nor in denial. It lies in patient, disciplined doubt — equally ready to say "yes, this is a concretion" and "no, this remains unexplained" — depending on what the facts show, not on what one wishes were true.

"Keep an open mind — but not so open that your brains fall out."

— attributed to James Oberg, aerospace engineer and skeptic

In the next chapter, we will change our direction of inquiry entirely. We will move away from individual objects — hammers, spheres, cubes, wedges — because, as this chapter has shown, individual objects almost never hold up to scrutiny across geological timescales. As we already learned in the second chapter, the planet destroys physical objects far too efficiently for us to build our evidence on them.

Instead, we will turn to what Gavin Schmidt and Adam Frank proposed in their famous 2018 paper: to search not for things, but for traces. Not hammers, but isotopes. Not cities, but chemical anomalies dissolved into the very fabric of the geological record — the places the geological eraser, for all its mercilessness, reaches last.

This is the "Silurian Hypothesis" — the most rigorous, the most scientifically grounded, and, strangely enough, the most unsettling question of all that this book has to ask.

✦ *A VOICE FROM THE STONE* ✦

Do not look for shape. Look for difference.

Shape disappears first — water eats it, fire eats it, time eats it. But difference lasts longer. The difference between what grows on its own, and what someone once guided.

We left you no hammers. We left you a question: why, in this layer of stone — this hair's-breadth strip of time — did something suddenly change? Why did the air turn different? Why did the earth, all at once, begin to breathe another way?

That is us. That is all that is left of us. Not a shape. Only the trace of a breath.

* * *

To be continued in Chapter Four: "The Silurian Hypothesis: The Science of Ghosts"

Chapter Four

The Silurian Hypothesis

A Science of Ghosts You Cannot See, But Can Weigh

In 2018, the academic journal International Journal of Astrobiology — a publication ordinary readers are unlikely to have ever come across on a bookstore shelf — published a paper with a deceptively modest title: "The Silurian Hypothesis: Would It Be Possible to Detect an Industrial Civilization in the Geological Record?"

Its authors — Gavin Schmidt, a climatologist who at the time headed NASA's Goddard Institute for Space Studies, and Adam Frank, an astrophysicist at the University of Rochester and one of the leading figures in the search for extraterrestrial life — were neither fringe figures nor sensation-seekers. These were scientists of impeccable reputation, publishing in the most serious journals, working at the intersection of climatology and astrobiology.

And yet they asked a question that had previously seemed to belong to the territory of tabloids and television specials about "ancient aliens": could a technological civilization have existed on Earth before humans — and if so, would we ever know it?

They borrowed the title from the British television series Doctor Who, where the Silurians were a fictional ancient reptilian race that had once ruled the Earth. The scientists chose the name with evident self-deprecating humor: they understood perfectly well how absurd such a question sounds coming from serious researchers.

But behind the self-irony lay an eminently serious scientific logic. And in this chapter, we will take it apart step by step.

✦✦✦

I. How Uncomfortable Questions Are Born

This hypothesis had at least two independent sources of inspiration — and both deserve separate attention, because they show how two seemingly unrelated scientific disciplines can unexpectedly converge on the same point.

Gavin Schmidt worked in climate modeling: reconstructing how the Earth's climate changed over millions of years, drawing on isotopic data extracted from sedimentary rock, ice cores, and marine sediments. A climatologist's work is inevitably bound up with one fundamental question: how do you tell natural climate fluctuations apart from fluctuations caused by the intervention of intelligent activity?

This question already confronts modern science when applied to ourselves. When climatologists study today's changing climate, they must rigorously, mathematically demonstrate that the observed warming is caused specifically by human activity — carbon dioxide emissions from burning fossil fuels — rather than by natural cycles the planet has gone through many times before, with no human involvement at all.

And here comes an unexpected twist: if science already knows how to distinguish an anthropogenic climate signal from a natural one in the present, could it not apply the same toolkit to the past? Could we search deep within the geological record for a similar signal — a climate anomaly bearing a distinctive "industrial" signature — belonging not to our own era, but to some other, far more ancient one?

Adam Frank, for his part, arrived at the same question from the opposite direction — from astrobiology and the famous Fermi Paradox: if the universe should be full of technological civilizations, why do we see no trace of any of them? Frank had long been interested in mathematically modeling how many civilizations might have arisen over the galaxy's entire history — and he insisted that, before drawing conclusions about the rarity of intelligence in the universe, we ought first to rigorously test how capable we even are of noticing intelligence, if it existed not on some other planet, but right here on Earth, simply a very long time ago.

Two questions — one about climate, the other about the cosmos — converged on a single point. And that point became the "Silurian Hypothesis."

"We asked this question not because we believe an ancient Earth civilization existed. We asked it because we realized science still has no clear answer for what its trace would look like — and without that answer, the very idea becomes untestable in either direction."

— from an interview with Gavin Schmidt, 2018

II. Why You Cannot Search for Bones

Before moving on, it's worth reinforcing the conclusion we already reached in the second chapter of this book: searching for physical objects — buildings, tools, skeletons — is essentially pointless on a geological timescale.

Schmidt and Frank state this as bluntly as possible in their paper. The probability that any given organism — let alone any given artificial object — will fossilize and survive to the present day is vanishingly small. And the probability that we would additionally happen to stumble upon that very specimen in the course of modern geological or archaeological work is smaller still, by several more orders of magnitude.

What's more: human civilization itself, for all its current numbers and ubiquity, will leave behind a remarkably meager set of skeletal remains in the geological record of the future. Most people today are buried or cremated in ways that essentially rule out fossilization. A hundred million years from now, if some alien geologist arrived to study our planet, the probability that they would stumble specifically upon a human skeleton, rather than the skeleton of some widespread animal, is statistically negligible.

If we ourselves, at the peak of our technological power and with a presence covering every landmass on the planet, are unlikely to leave behind discoverable bones — what makes us think a hypothetical ancient civilization, smaller in number and far more remote in time, would have left any?

This is exactly why Schmidt and Frank set aside, from the very outset, any search for skeletons, ruins, or artifacts — the very objects we examined in the third chapter. Instead, they propose searching not for "things," but for "chemistry." Not for shape, but for concentration.

✦ ✦ ✦

III. Carbon: Deep Time's Master Storyteller

At the heart of the Silurian Hypothesis's methodology lies one of the most powerful tools in modern geochemistry: the analysis of carbon isotope ratios in sedimentary rock.

Let's put it simply. Carbon exists in nature as several isotopes — atoms with the same number of protons but a different number of neutrons. The two main stable isotopes are carbon-12 and carbon-13. Their ratio in rocks, fossils, and ice cores carries information about where that carbon came from.

The key fact is this: living organisms, during photosynthesis, preferentially absorb the lighter isotope — carbon-12. This means that carbon which has passed through biological processes — carbon accumulated in ancient plants, say, and later turned into coal or oil — is systematically "depleted" in the heavier carbon-13 isotope, compared to carbon that never took part in life.

When humanity began burning fossil fuels on a massive scale — coal, oil, gas — it was, in effect, injecting into the atmosphere enormous volumes of this "biologically light" carbon, accumulated over millions of years within the earth.

The result is already clearly visible in geochemical data: since the start of the Industrial Revolution, the ratio of carbon isotopes in the atmosphere has shifted noticeably — specialists call this the "Suess effect," after the Austrian scientist Hans Suess, who first described the phenomenon back in the mid-twentieth century.

This is exactly the "industrial signature" climatologists know how to recognize in modern data. And it is precisely this signature that Schmidt and Frank propose searching for deep within the geological record.

HOW TO READ THE ISOTOPE TRACE

Carbon-12 — the "light" isotope, preferentially absorbed by living organisms

Carbon-13 — the "heavy" isotope, less often drawn into biological processes

Burning fossil fuels releases carbon into the atmosphere "depleted" in carbon-13 — the so-called Suess effect

This same ratio is recorded in sedimentary rock, corals, and tree rings — and could potentially be found in ancient deposits of any age

IV. The PETM: A Catastrophe Nature Rehearsed All on Its Own

To understand exactly what to look for, Schmidt and Frank turn to an event that nature itself once staged without any involvement of intelligent beings — and one strikingly similar, in its geochemical profile, to what humanity has done over the last two centuries.

This event is called the Paleocene-Eocene Thermal Maximum, or PETM for short — a razor-sharp episode of climate warming that occurred roughly fifty-six million years ago, at the boundary between two geological epochs.

Over a geologically negligible span of time — variously estimated at anywhere from a few thousand to perhaps twenty thousand years, which is nearly an instant in geological terms — a colossal quantity of carbon depleted in carbon-13 was released into Earth's atmosphere and oceans, matching almost exactly, in isotopic profile, what is happening today because of our own burning of fossil fuels.

The result was a sharp, dramatic rise in global temperature — by various estimates, five to eight degrees Celsius — ocean acidification, a mass extinction of deep-sea organisms, and a radical reorganization of ecosystems across the entire planet. The trace of this event survives in sedimentary rock around the world as a distinctive, easily recognizable "negative excursion" in carbon isotopes — a sharp, near-instantaneous shift in their ratio that any geochemist can identify almost at a glance.

The cause of the PETM remains a subject of scientific debate to this day. The leading hypotheses point to natural, purely geological sources: large-scale volcanism, methane release from ocean-floor sediments, the destabilization of methane hydrates triggered by other climate shifts. No serious scientist proposes that the PETM was caused by intelligent activity — the event fits entirely within the bounds of natural geological processes.

But this is exactly why the PETM matters so much to the Silurian Hypothesis: it serves as a kind of benchmark, a "control sample" for what a sharp, large-scale injection of light carbon into the atmosphere looks like in the geological record. If an ancient technological civilization had burned fossil fuels the way we do, the trace it left behind, by Schmidt and Frank's logic, should look geochemically very much like the PETM.

"If somewhere in the geological record there exists yet another sharp, unexplained isotopic excursion, similar in profile to the PETM but lacking an obvious geological trigger — that would be the first place worth a closer look."

— the logic of Schmidt and Frank's paper, paraphrased

V. The Plastic That Will Outlive Us: Technostratigraphy

Carbon isotopes are not the only possible trace. Schmidt and Frank, and following them, other researchers studying the geology of the Anthropocene, point to a whole set of additional markers that together form what is now commonly called "technostratigraphy" — the study of crustal layers bearing the imprint of a technological civilization.

The most durable of these markers is, strangely enough, plastic.

Synthetic polymers — plastic, in the sense we're familiar with — are molecular structures that essentially never occur in nature apart from deliberate, purposeful activity. Complex, long polymer chains with artificial bonds, stable additives, specific molecular "fingerprints" — all of this is remarkably resistant to ordinary biological decay. Some studies indicate that certain kinds of plastic can survive in anaerobic, oxygen-free sedimentary conditions — on the floor of deep ocean trenches, for instance — for geologically significant stretches of time, potentially millions of years, before they are either buried deep enough to undergo metamorphism or destroyed by other processes.

Scientists have already found modern plastic fused into natural geological formations — so-called "plastiglomerates," in which plastic melted by fires or volcanic activity mixes with sand, shells, and coral and hardens into a single rock-like mass. Such formations were first documented in Hawaii in 2014 by the geologist Patricia Corcoran.

If a hypothetical ancient civilization had also used synthetic polymers — even at a chemical level fundamentally different from our own — the distinctive molecular structure of such compounds could, in theory, survive hundreds of millions of years under sufficiently protected sedimentary conditions, even as any physical objects made from that same material would long since have been destroyed.

VI. Metals That Don't Occur in Nature — and Radioactive Ash

Another class of potential markers is rare metals and alloys that don't occur in nature in pure form or in certain proportions.

Modern industrial civilization actively uses and redistributes across the planet a great many elements that, under natural conditions, are either extremely rare or evenly dispersed at very low concentrations: rare-earth metals used in electronics, specific alloys not produced by any known geological process. A sharp, anomalous concentration of such elements within one narrow geological layer, repeated across different continents simultaneously, would constitute a serious signal.

But perhaps the most specific — and most unambiguous — marker Schmidt and Frank propose has to do with nuclear technology.

Nuclear weapons tests and the operation of nuclear reactors leave a distinctive radioactive trace — in particular, the isotope plutonium-239, which essentially does not exist in nature under natural conditions and forms almost exclusively as a result of artificial nuclear reactions. Mid-twentieth-century nuclear weapons tests have already left a thin but utterly unambiguous layer of this isotope in global geological deposits — in ice cores, in lake-bottom sediments — one that many geologists propose using as the official marker for the start of the Anthropocene, a new geological epoch defined by the presence of humanity.

Plutonium-239 has a half-life of roughly twenty-four thousand years — which makes it useless for searching for events hundreds of millions of years old: over such a span, the isotope would decay completely, leaving no measurable trace. But the underlying principle — that nuclear activity leaves a distinctive, easily identifiable isotopic signature, fundamentally different from any natural process — remains critically important for understanding what a "fingerprint" of intelligent technological activity might, in principle, look like.

If a hypothetical ancient civilization had developed nuclear power, corresponding isotopes with longer half-lives could, in theory, still carry a measurable trace even after hundreds of millions of years. Though, in fairness, it must be said: no geological layer known today shows anything of the kind.

✦ *A VOICE FROM THE STONE* ✦

We left behind no ash you would recognize as the ash of war.

Our fire was different — slow, alive, growing, not devouring. We did not tear open the earth to reach what lay hidden beneath it. We asked. Sometimes the earth answered. Sometimes it did not.

If you find us, you will not find an explosion. You will find a slow change — air grown slightly different over many generations, a sea whose taste shifted just a little, a forest that began growing not quite the way it grew before.

Do not look for a scar. Look for the quiet shift in the planet's breathing.

VII. A Needle in a Haystack the Size of a Planet

Everything discussed so far sounds reassuringly methodical: there are clear, measurable, physically grounded markers one could, in principle, search for. But here we must, in all honesty, pause on a colossal practical problem that Schmidt and Frank themselves acknowledge openly.

The geological record available for study is extremely incomplete and extremely unevenly distributed across the planet. Outcrops of rock of the right age — places where the geological layer that interests us is physically accessible at the surface for sampling, rather than buried at an unreachable depth or, as we learned in the second chapter, simply not preserved at all — make up a vanishingly small fraction of Earth's surface.

What's more: even where the right layer is accessible, studying it requires a targeted, deliberate search for precisely the type of anomaly we're looking for. And scientists, for perfectly understandable reasons, do not spend their time and resources hunting for a hypothetical signal of civilization in every sedimentary section they study — they are looking for answers to far more down-to-earth, testable questions: climate reconstructions, the history of extinctions, the evolution of particular groups of organisms.

This means that even if such a signal really does exist somewhere in the Earth's crust, the probability that it has already been noticed by chance and correctly interpreted is extremely low. We could literally be holding the right specimen in our hands and not know it, because we were looking for something else entirely.

Schmidt and Frank state this problem without any theatrics: the absence of a detected signal is not proof that no signal exists. It could equally be proof that we never truly looked for it.

"We are not engaged in searching for proof that an ancient civilization existed. We are engaged in an honest assessment of how capable our science even is of answering this question — in either direction."

— Gavin Schmidt and Adam Frank, "The Silurian Hypothesis," 2018

✦✦✦

VIII. Caution as a Scientific Virtue

A serious methodological caveat is needed here — and it would be dishonest to avoid it.

The PETM, which we discussed in Section IV, is far from the only sharp isotopic excursion in the planet's history. The geological record contains dozens of similar events across hundreds of millions of years: climate anomalies tied to volcanism, to shifts in ocean currents, to fluctuations in Earth's orbit, to massive methane releases from seafloor sediments during natural warming. The overwhelming majority of such events have a fully convincing, thoroughly studied natural explanation that requires no recourse to a hypothesis of intelligent activity.

This is critically important to understand: the mere fact of discovering a sharp isotopic shift in ancient rock is not, by itself, proof that an ancient civilization existed. That would be exactly the kind of crude logical error the entire third chapter of this book warns against: hastily declaring any anomaly to be proof of whatever we want to believe.

Schmidt and Frank state this directly and insistently in their paper: the correct conclusion to draw from discovering an unexplained isotopic anomaly is not "we found a civilization," but "we need more thorough investigation to rule out every known natural cause before considering any other hypothesis." The vast majority of such anomalies will, in all likelihood, turn out on close examination to have a perfectly mundane geological explanation — just as happened with nearly every artifact in the third chapter.

The value of the Silurian Hypothesis lies not in asserting that an ancient civilization existed. Its value lies in finally formulating a rigorous, testable methodology for answering this question honestly — a methodology far more likely to prove we will find nothing than the reverse.

And this, paradoxically, is a far more serious contribution to science than any sensational find. Because before this paper, the question of an ancient Earth civilization existed outside science entirely — a realm of speculation, immune to both proof and disproof. After it, the question became part of the normal scientific agenda: a formulated hypothesis, specific predictions, specific methods of testing.

✦ ✦ ✦

IX. A Science of Ghosts

There is a peculiar, unsettling beauty in the idea of a "science of ghosts" — a discipline that searches not for a body, but for a shadow; not for an object, but for its absence, recorded in the chemistry of the planet.

This is an unusual stance for a science accustomed to working with physical specimens, with things you can touch, weigh, photograph under a microscope. The Silurian Hypothesis asks us to search for something far more ephemeral: a shift in isotope ratios, dissolved within a body of sedimentary rock — a statistical ghost of an activity that perhaps never left behind anything more tangible than this.

And yet it is precisely this ephemerality that makes the hypothesis so powerful. Because it does not ask: "Have we seen a city?" It asks: "Did the planet breathe differently at some point in its history — the way it began breathing differently after we built our first factories?"

This is a question that can, in principle, be answered — provided one searches for it correctly and patiently. And this is exactly why the Silurian Hypothesis is neither a sensation nor a hoax. It is an invitation to the most difficult and the most honest kind of scientific inquiry: the search for something that perhaps never existed at all — but whose absence we cannot yet rigorously prove.

"We stand at the threshold of a new discipline — a kind of forensic geochemistry on a planetary scale, where the evidence is not fingerprints, but shifts in the atomic weight of carbon."

— an idea paraphrased from subsequent scientific commentary on the 2018 paper

In the next chapter, we will finally take up the question we have so carefully set aside until now: if such a civilization existed, who might have created it? Which creatures, long before mammals ever appeared, possessed the biological preconditions for developing intelligence and technology — and along what fundamentally non-human paths might their ingenuity have traveled?

We will set out in search of candidates for a throne that may have stood empty long before any of our own ancestors could ever have laid claim to it.

* * *

To be continued in Chapter Five: "Intelligence Before Humanity: Candidates for the Throne"

Chapter Five

Intelligence Before Humanity

Candidates for a Throne They Never Knew Existed

Close your eyes and picture an intelligent being.

Chances are, the image that came to mind had two arms, two eyes, an upright posture, perhaps a large, domed forehead concealing a sizable brain. Even if you try to picture something "alien," you will most likely still endow it with the symmetry of our own bodies, because we don't really have any other language for thinking about intelligence.

This is not a failure of imagination. It is the inevitable consequence of the fact that we are the only known example of intelligent life, and our brains are forced to extrapolate the concept of "intelligence" from the single data point available to them: themselves.

But what if intelligence is under no obligation to look like us?

What if it can grow inside a body built on entirely different engineering principles — with an external skeleton instead of an internal one, with a dozen eyes instead of two, with a distributed nervous system instead of a centralized brain? What if the technology such a creature would produce were so alien to our expectations that, stumbling upon its traces, we simply wouldn't recognize them as technology at all?

In this chapter, we set out in search of candidates — creatures that lived on Earth hundreds of millions of years before the first primate, which, in theory, possessed the biological preconditions for developing something we might call intelligence. We are not claiming any of them actually built a civilization. We are asking a more modest, but no less imagination-stirring question: if it had happened — who would be the most likely contenders, and what would their intelligence have looked like?

✦✦✦

I. What It Actually Takes to Become Intelligent

Before searching for candidates, we need to define our criteria. And here the first difficulty arises: science still has no single, universally accepted definition of intelligence.

Even so, we can identify a set of biological preconditions that, in the view of most evolutionary biologists, significantly raise the odds of a technological civilization arising. There are several of them.

First — a developed central nervous system capable of complex information processing and, ideally, of abstract thought. Not merely a bundle of reflexes, but a system capable of building internal models of the world and predicting the consequences of its own actions.

Second — manipulators: limbs or other organs precise and dexterous enough to interact with the environment at a fine-grained level — grasping, shaping, working materials.

Intelligence without "hands" — in whatever form they might take — risks remaining an observer's intelligence, unable to materialize its ideas as technology.

Third — sociality: the ability to pass accumulated knowledge on to the next generation, to coordinate complex collective action. A solitary genius, however brilliant, can rarely build a civilization alone. Technology is almost always the product of knowledge accumulated and transmitted across many generations.

Fourth — a sufficient lifespan and a stable enough habitat to allow learning and the accumulation of experience over an individual's lifetime, rather than acting purely on rigidly fixed instinct.

Notice: not one of these criteria specifically requires two hands, an upright posture, or mammalian origins. These are functional requirements that evolution can satisfy through any number of different paths, provided environmental conditions push it in that direction.

"Evolution is not an architect working from a single blueprint. It is millions of independent experiments run in parallel, only a handful of which have ever produced anything resembling intelligence as we know it."

— a widely held principle of evolutionary biology

II. The Carboniferous Giants: A World Where Insects Were Kings

We already spoke, in the first chapter, about the remarkable atmosphere of the Carboniferous — saturated with oxygen up to thirty-five percent, against today's twenty-one. This detail is not merely a curious fact. It explains what is perhaps the single most striking biological phenomenon of that era: arthropod gigantism.

Meganeura, a giant dragonfly whose fossils are found in French coal seams, had a wingspan of up to seventy centimeters — comparable to a large bird of prey. Arthropleura, a giant millipede, grew to two and a half meters in length — the largest terrestrial arthropod known to science in the planet's entire history. There were spiders the size of a dinner plate, scorpions as long as a human arm.

The cause of this gigantism lies in a peculiarity of the arthropod respiratory system: they breathe not through lungs, but through a system of tracheae — extremely fine tubes that deliver oxygen directly to tissues without the involvement of blood. This system is efficient, but it has a physical limit: the larger the body, the harder it is for oxygen to diffuse deep along the tracheae. This is exactly why, in today's less oxygen-rich atmosphere, insects are confined to comparatively modest sizes. But at an oxygen concentration of thirty-five percent, that physical ceiling shifted considerably higher, allowing arthropods to reach sizes that seem almost fantastical today.

A tempting question arises: if the body could grow this large, could the nervous system have grown right along with it — all the way to something resembling complex intelligence?

GIANTS OF THE CARBONIFEROUS

Meganeura monyi — a dragonfly with a wingspan of up to 70 cm

Arthropleura — a millipede up to 2.5 meters long, the largest known terrestrial arthropod

Pulmonoscorpius — a scorpion roughly 70 cm long

Atmospheric oxygen concentration: up to 35% (today's level: 21%)

Period: roughly 359–299 million years ago

✦ ✦ ✦

III. The Exoskeleton Problem: Why Gigantism Is Not the Same as Intelligence

Here we must cool our imagination with some sober biological reality.

The arthropod nervous system is built on fundamentally different principles from that of vertebrates. Instead of a single centralized brain, insects and their relatives have a distributed system of nerve clusters — ganglia — each capable of running its own body segment with relative autonomy. This makes their behavior extremely effective for fast, rigidly programmed reactions, but poorly suited to the kind of flexible, abstract thought we associate with intelligence.

There is also a more fundamental physical constraint. The exoskeleton — the external skeleton characteristic of arthropods — provides superb protection, but imposes a hard limit on brain growth. To grow, an arthropod must periodically molt — shedding its old shell and, during a brief, extremely vulnerable window, building a new, larger one. A brain enclosed in the rigid capsule of an exoskeleton cannot expand as smoothly and continuously as the brain of a vertebrate, enclosed within a gradually growing skull.

What's more, the larger an arthropod becomes, the more catastrophic the molting process is for it: a gigantic exoskeleton demands gigantic expenditures of calcium and chitin with every molt, and the vulnerable window after shedding the old shell, before the new one hardens, becomes lethally dangerous for an organism of that size.

In other words: the exoskeleton brilliantly solves the problem of physical protection and allows impressive sizes to be reached in an oxygen-rich atmosphere, but it also, evidently, sets an evolutionary ceiling on the path toward a complex, centralized brain that keeps growing throughout an individual's life — the very neural architecture that, as far as we know from the example of vertebrates, is best suited to abstract thought.

This does not mean intelligence built on an arthropod architecture is categorically impossible in principle — evolution is rich in unexpected solutions, and some modern insects, social ones in particular, display astonishingly complex collective behavior (we'll discuss this separately, a bit later in this chapter). But individual, "human-like" intelligence in a single arthropod — given the biological constraints we know of — looks like an evolutionarily improbable path.

✦ *A VOICE FROM THE STONE* ✦

We do not remember ourselves the way you remember yourselves.

None of us had a single center where all our thoughts lived at once. A thought was born wherever it was needed — in the front segment, when a decision had to be made; in the rear, when escape was needed. We were a multitude of small decisions, harmonized into a single whole, rather than one great will governing the body from above.

You would call this a lack of mind. We called it something else: we were agreement, not command.

Perhaps you are right that we did not build what you will build. But we lived in this forest longer than you will live. That, too, is a kind of skill.

IV. Giant Amphibians: The Tetrapods of the Carboniferous

If arthropods run into a hard biological ceiling, another group of creatures from the same era looks like a far more promising candidate — the early tetrapods, four-limbed vertebrates that included the ancestors of modern amphibians and reptiles.

It was precisely during the Carboniferous that this group experienced a genuine flowering of diversity. There were giant amphibious predators — such as the temnospondyls, some of which reached several meters in length and looked something like a cross between a crocodile and a salamander. A later, but telling example is Prionosuchus, which lived in the following Permian period and reached up to nine meters in length, making it one of the largest amphibians in the entire history of the Earth.

Unlike arthropods, tetrapods have an internal skeleton and a centralized nervous system with a full-fledged brain enclosed in a skull that grows along with the animal throughout its life — the same basic architecture that later allowed the mammalian brain, including our own, to develop. This is a fundamentally different evolutionary trajectory, and a far more promising one from the standpoint of intelligence's potential development.

Early tetrapods already had limbs capable, given sufficient evolutionary pressure, of specializing into manipulators — as ultimately happened along the line leading to mammals, and eventually to primates. They had a developed sensory system, and a capacity for learning of the kind found in many modern amphibians and reptiles — though, of course, none of them ever came close to the level of abstract thought characteristic of humans.

The question that naturally arises here is: why didn't any lineage of Carboniferous amphibians or early reptiles follow the path toward a large brain and complex social behavior — the same path primates would follow millions of years later?

The answer, it seems, has less to do with any fundamental impossibility than with the absence of sufficient evolutionary pressure in that direction. Developing a large, energy-hungry brain is an extraordinarily costly evolutionary choice: in modern humans, the brain consumes roughly twenty percent of the body's total energy at rest, despite making up only about two percent of body mass. Evolution only "chooses" such costly solutions when specific environmental conditions make them advantageous — complex social competition, say, the need to solve non-routine problems in order to find food, or an extended period of parental care.

For the early tetrapods of the Carboniferous, this kind of specific pressure evidently just never arose — or at least, never arose strongly or persistently enough to trigger the same evolutionary brain arms race that, almost three hundred million years later, gave rise to humans.

"A large brain is not a reward for long evolution. It is a costly response to a very specific set of ecological circumstances, one that evolution stumbles upon only rarely."

— a widely held principle of evolutionary neuroscience

✦ ✦ ✦

V. Cephalopods: An Intelligence Already Here — and One That Will Never Fossilize

Before moving further back into deep time, it's worth taking a curious detour — because it will help us understand something important about how intelligence can hide from the geological record even today, right before our eyes.

Octopuses and other cephalopods are among the most cognitively sophisticated creatures on the planet, with brains that share essentially no evolutionary kinship with the vertebrate brain. Their lineage diverged from the one leading to humans roughly six hundred million years ago — before complex nervous systems, in the form we're familiar with, even existed. Octopus intelligence evolved entirely independently, along a completely different evolutionary path.

And this intelligence is genuinely alien. An octopus has no single centralized brain in our sense of the term — a substantial share of its neurons, roughly two-thirds, are distributed throughout its arms, each capable of making its own, semi-autonomous decisions. Octopuses have demonstrated the ability to solve complex puzzles, to use simple tools — hiding inside a coconut shell, for instance — to recognize individual people, and to display something resembling individual curiosity and playful behavior.

And yet, for all that, cephalopods almost never fossilize in any recognizable form. They have no hard skeleton — only soft tissue, which decomposes leaving virtually no trace under the vast majority of geological conditions. The only thing that usually survives of ancient cephalopods in the fossil record is the shells of their distant ancestors, ammonites and the like

— but the soft-bodied, highly intelligent modern forms themselves leave behind almost no recognizable fossils at all.

Consider this: right now, in the oceans of the modern Earth, there exists an evolutionarily independent form of genuine, full-fledged, sophisticated intelligence — and if humanity vanished tomorrow, future geologists studying our era a hundred million years from now would most likely never learn that octopuses were intelligent at all. Almost nothing would remain of them, apart from extraordinarily rare, chance impressions under exceptionally favorable conditions.

This is a critically important illustration for the entire theme of this book: intelligence and fossilizability are two completely independent properties. A creature can be extraordinarily intelligent and still leave behind a geological zero. And if this is possible for creatures living right now, right next to us, what would have prevented the exact same scenario from playing out three hundred million years ago, with some soft-bodied or poorly preserved creature of the Carboniferous whose existence we may never learn of at all?

THE OCTOPUS AS A LESSON IN HUMILITY

The octopus brain evolved independently of the vertebrate brain — the common ancestor lived roughly 600 million years ago

Roughly two-thirds of an octopus's neurons are located not in the central brain, but in its arms

Octopuses have demonstrated simple tool use, puzzle-solving, and individual recognition of people

Soft-bodied organisms are almost never preserved in the geological record — fossilization requires hard tissue

Conclusion: intelligence and the capacity to leave a fossil are independent of one another

✦✦✦

VI. Collective Intelligence: Technology Without a Genius

There is one more model of intelligence we almost never take seriously, because it doesn't fit our intuitive picture of an "intelligent being" as a single, thinking individual.

Social insects — ants, termites, some bees — display behavior entomologists call "swarm intelligence," or superorganism intelligence. A single ant has extremely modest cognitive abilities and acts mostly on simple, rigidly fixed rules. But an ant colony as a whole — as a unified system — is capable of solving problems of striking complexity: building ventilated, multi-level cities out of earth; practicing agriculture (some ant species have cultivated fungus for millions of years); waging wars with clear military strategy; optimizing supply logistics no worse, and sometimes better, than algorithms designed by humans.

The termite mounds of certain species reach several meters in height and constitute extraordinarily sophisticated engineering structures, complete with a natural ventilation system capable of maintaining stable temperature and humidity inside the colony regardless of fluctuations outside — a kind of natural air conditioner, designed not by a single architect, but by millions of years of collective behavioral evolution.

Termites, which is especially important to our theme, are an extraordinarily ancient group of insects, whose ancestors can be traced in the fossil record all the way back to the Carboniferous and Permian periods. This means the model of collective, distributed intelligence — embodied not in a single large individual, but in the coordinated behavior of countless small ones — already existed on Earth hundreds of millions of years ago, and, in essence, still exists today.

What if a technological civilization does not necessarily require a single genius with a large brain? What if it could arise — slowly, without any single creative flash of insight, but with the same inexorability with which a termite mound grows into a sophisticated engineering structure — as a byproduct of the collective, distributed behavior of a great many relatively simple organisms?

Of course, there is a colossal gulf between a termite mound and a city — termites do not invent, they make no qualitative leaps in their understanding of the world, and their architecture is genetically fixed rather than recreated anew by each generation on the basis of accumulated, culturally transmitted knowledge. But the idea itself matters: intelligence and technology need not be tethered to a single, large, centralized nervous system. There exists, at least in theory, another path — the path of the swarm.

✦ *A VOICE FROM THE STONE* ✦

No single one of us, alone, was wise.

Wisdom did not live in one head — it lived between us, in how we passed a signal, a scent, a movement, from one to another. Not one of us understood the whole design. But the design was carried out all the same — the way breathing happens without asking permission of any single cell in the lung.

You will look for one of us — a genius, a leader, a maker. You will not find one. We were never a single being. We were always many, acting as one.

Perhaps that is why you will mistake us for an anthill. Perhaps you will be right — and still be wrong.

VII. What Their Technology Might Have Looked Like

If we allow — purely hypothetically — that one of these candidates, or something like them, did in fact cross the threshold separating complex behavior from genuine technology, it's natural to ask: what would that technology have looked like? Would we even recognize it as technology at all?

Here we step into the territory of cautious, frankly speculative extrapolation — and we must flag this as clearly as possible. But the extrapolation itself is instructive, because it forces us to question one of our most deeply rooted assumptions: the idea that technology necessarily means metal, fire, and the wheel.

The technology of social insects, even today, is predominantly a technology of architecture and chemistry, not metal and mechanics. Termites don't forge tools — they grow ventilated structures. Ants don't invent writing — they encode astonishingly complex information in chemical pheromone trails, capable of conveying data about routes, threats, and food quality with a precision that some human communication systems might envy.

If we push this logic several orders of magnitude further — imagining a hypothetical civilization that grew out of just such a collective, chemical-architectural tradition — its technology might have been built around fundamentally different foundations: biological engineering instead of metallurgy, the deliberate breeding and modification of other species instead of machine-building, intricate chemical systems for conveying information instead of writing in the sense we're familiar with.

Such a technology would leave behind neither gears nor skyscraper foundations. It would leave behind, perhaps, only strangely organized ecosystems, inexplicably stable symbiotic relationships between species, local anomalies in soil chemistry — precisely the sort of thing a modern geologist or biologist, stumbling upon such a trace in the fossil record, would most likely write off as "just an unusual natural system," without ever suspecting any design behind it.

"We search for mirrors. We search for ourselves in the past. But if an intelligence before us was alien, it would leave behind no mirror — it would leave behind something we look into and see only scenery."

— an idea paraphrased from the philosophy of recognizing non-human intelligence

✦✦✦

VIII. Caution Against Imagination

It's important to state as plainly as possible: nothing listed in this chapter is proof that any of these creatures — giant arthropods, early tetrapods, social insects — actually developed a technological civilization. The paleontological record contains not a single convincing piece of evidence, to date, of such a development in any of these groups.

What's more, every candidate discussed here faces serious biological constraints that modern science considers substantial obstacles on the path to a technological civilization in any recognizable sense: a hard ceiling on brain growth in arthropods, insufficient evolutionary pressure toward a large brain in early tetrapods, and behavior fixed genetically rather than culturally in social insects, which makes a qualitative leap in development highly improbable without prolonged, directed selection specifically toward that end.

But the purpose of this chapter is not to prove that an ancient intelligence existed. The purpose is to widen the space of what we're even capable of imagining as "intelligence" and "technology," before we continue searching for their traces in the remaining chapters of this book. Because if we search only for a two-handed, upright, large-brained creature building with stone and metal, we will probably miss any genuinely alien form of intelligence — even one that left behind a far more noticeable trace than we expect, simply not in a form we've been trained to recognize.

The mirror we look into, searching for ancient intelligence, mainly reflects ourselves. And that is the single greatest methodological danger in this book's entire subject.

IX. A Mirror That Reflects Nothing

There is a peculiar irony in the fact that the most promising candidates, from an evolutionary standpoint, for intelligence before humanity are not the giant, spectacular creatures that dazzle the imagination with their size, like Meganeura or Arthropleura. The most promising candidates are the modest, centrally organized tetrapods with a brain growing inside a skull — whose lineage, after nearly three hundred million more years of evolution, eventually produced us.

Perhaps the only truly honest conclusion of this chapter runs as follows: the path evolution ultimately took toward humanity was a rare, narrow, almost accidental route through a vast space of evolutionary possibility. The Carboniferous forests were full of experiments — some ran into dead ends of gigantism, others into dead ends of narrow specialization, still others into models of collective, distributed behavior. None of them, as far as we know, led to anything resembling a technological civilization.

But "as far as we know" is not the same as "for certain." And if the previous chapters have taught us anything, it's that the geological record is too incomplete, too selective, too easily erased by the geological eraser for us to close this question with complete confidence.

"The absence of recognizable intelligence in the record is not the same as the absence of intelligence. Perhaps we simply don't know what the mirror worth looking into actually looks like."

— a closing thought for this chapter

In the next chapter, we will turn to ourselves — to the one species whose technological nature is beyond doubt. We will ask: what will remain of our own civilization a hundred, two hundred, three hundred million years from now? What trace — plastic, radioactive, isotopic — will outlive us, in a way nothing outlived the hypothetical civilizations of the Carboniferous?

Because perhaps the best way to understand what to look for in the past is to first look honestly into the mirror of the present.

* * *

To be continued in Chapter Six: "The Shadow of Plastic and Radiation"

Chapter Six

The Shadow of Plastic and Radiation

What We Will Leave Behind — Whether We Want To or Not

In the previous chapter, we searched for intelligence deep within the geological past, among giant dragonflies and collective insects. Now it's time to turn the telescope the other way — and point it not at the past, but at the future.

Picture a geologist living a hundred million years from now. Not a human — it's hard even to guess what kind of creature that would be, if any exists at all. Simply an intelligent observer, armed with methods resembling our own, who one day stumbles upon a strange, thin layer within a body of sedimentary rock — a layer unlike anything lying above or below it.

What would they find there? Which of the things we've created is truly capable of surviving a hundred, two hundred, three hundred million years — the very span we discussed in connection with hypothetical civilizations of the Carboniferous?

This question is not an idle exercise in imagination. It is the only way available to us to test the entire logic of the previous chapters against a known sample. We are the one technological civilization whose existence is beyond doubt. If we understand what will outlive us specifically, we will gain a far more honest sense of what's worth searching for when asking about someone who came before us.

✦✦✦

I. The Anthropocene: An Epoch in Search of Its Own Beginning

Geology has a strict, almost bureaucratic procedure for recognizing a new epoch. It's not enough simply to declare, "we live in a new age." The International Commission on Stratigraphy requires a so-called "golden spike" — a Global Boundary Stratotype Section and Point, a specific physical location within a specific geological section where the transition from one epoch to another is recorded clearly, unambiguously, and universally.

Since the early 2000s, geologists have been engaged in a lively, at times fierce debate: does humanity's impact on the planet warrant designation as a distinct geological epoch — the Anthropocene — and if so, exactly where does that boundary lie?

Various candidates were considered: the beginning of agriculture roughly twelve thousand years ago, the start of the Industrial Revolution at the end of the eighteenth century, the mid-twentieth century. In the end, a special working group that studied this question for more than a decade settled on the sharpest, most unambiguous marker available: the mid-twentieth century, specifically the period around

1950–1952, when large-scale thermonuclear weapons testing began.

Why exactly this? Because that was the moment when, unlike the slow, gradual changes of climate or land use, the planet received an instantaneous, global, utterly unambiguous signal — a thin layer of the radioactive isotope plutonium-239, recorded almost simultaneously in the ice cores of Greenland and Antarctica, in lake-bottom sediments on different continents, and in the corals of tropical seas.

It is telling that the commission of geologists, faced with a choice among dozens of possible indicators of humanity's footprint on the planet, settled specifically on radioactivity — not plastic, not concrete, not climate change, but precisely the marker that points, absolutely unambiguously, with no room for alternative reading, to technological rather than merely biological activity.

"The mid-twentieth-century plutonium layer is probably the sharpest, most unambiguous geological marker in the planet's history over the last several million years. Sharper than most markers of mass extinctions."

— from discussions of the Anthropocene Working Group

II. The Planet's Plastic Skin

But plutonium, as we already noted in the fourth chapter, decays comparatively quickly by geological standards — a half-life of roughly twenty-four thousand years means that within just a few million years, this particular marker will become indistinguishable. If we want to understand what can genuinely survive a hundred million years or more, we need to look for something more durable.

Here, a material moves to the forefront that, in everyday thinking, is associated more with pollution and garbage than with historical grandeur — plastic.

We already mentioned, in the fourth chapter, the phenomenon of "plastiglomerates" — hybrid rocks in which molten plastic mixes with sand, shells, and coral and hardens into a single, literally stone-like mass. But the scale of the planet's plastic pollution extends far beyond a handful of exotic finds on Hawaiian beaches.

By various research estimates, humanity has produced more than eight billion tons of plastic since the mid-twentieth century. The overwhelming majority of this volume — over seventy-five percent — has already become waste. A substantial share of this waste ultimately ends up in the ocean, where currents carry it enormous distances, and part of it inevitably settles on the seafloor, above all in deep ocean trenches — precisely the zones which, as we know from the second chapter, are among the most geologically stable regions of the planet.

Microplastic — tiny fragments of degraded polymers smaller than five millimeters — has already been found virtually everywhere: in the seafloor sediments of the deepest points of the world's oceans, including the Mariana Trench, in Arctic ice cores, in the soils of the remotest mountain regions where no human foot has ever set down. Some studies indicate that microplastic, under certain anaerobic, oxygen-poor seafloor conditions, can persist for geologically significant periods of time, potentially measured in millions of years, before either fully breaking down or, more likely, becoming buried deep enough to undergo metamorphism.

Plastic becomes a geological marker not because it is literally eternal, but because it is chemically alien: long artificial polymer chains, specific molecular structures, certain kinds of stabilizing additives — none of this is produced by any known natural, non-biological process. Even after breaking down to the molecular level, the distinctive chemical "fingerprint" of synthetics can persist within sedimentary rock far longer than the recognizable shape of the object itself.

THE SCALE OF THE PLASTIC FOOTPRINT

Over 8 billion tons — total plastic produced by humanity since the mid-20th century

Over 75% of this volume has already become waste

Microplastic has been found in the seafloor sediments of the Mariana Trench, the deepest point of the world's oceans

Plastiglomerates were first documented in Hawaii in 2014 by the geologist Patricia Corcoran

✦ ✦ ✦

III. Concrete: The Mountain We Built

If plastic is the most chemically alien of our materials, concrete is perhaps the most massive.

The total mass of all concrete ever produced by humanity, by various estimates, approaches five hundred billion tons — a figure of the same order of magnitude as the mass of all plant biomass on the planet. This makes concrete one of the most substantial materials, by volume, ever synthesized on Earth by any living thing, rivaling the combined activity of the entire biosphere over millions of years.

Concrete, unlike plastic, is not, in the strict sense, a purely synthetic molecular structure — it is, in essence, an artificial conglomerate binding natural materials (sand, gravel) together with artificially manufactured cement. This is exactly why, counterintuitively, individual concrete structures are subject to relatively rapid decay by geological standards: cement gradually carbonates, loses strength, and the structure crumbles under the action of water, temperature swings, and plant roots. Most modern buildings, left unmaintained, will crumble beyond recognition within a mere few centuries, a millennium or two at most.

However, the mere presence of concrete — even in the form of shattered, scattered rubble — leaves behind a specific geochemical trace. Cement production radically alters the local calcium balance and locks away carbon dioxide in a distinctive way, through a process called carbonation — a process that leaves behind a recognizable isotopic and mineralogical signature, different from that produced by the natural weathering of rock.

If plastic and radioactive isotopes are bright, unambiguous signals, concrete will most likely survive in the geological record of the future more as a subtle, but statistically significant, shift in the chemistry of sedimentary rock beneath major urban agglomerations — a kind of broad, blurry smudge rather than a sharp signal.

IV. Metals That Don't Occur in Nature

There is one more category of the human footprint worth singling out — metal alloys and compounds that physically do not form through any known natural geological process.

Stainless steel — an alloy of iron with chromium and nickel in strictly defined proportions — occurs in nature in no form whatsoever: a combination of elements like this, at concentrations like this, requires deliberate, controlled metallurgical processing at precisely regulated temperatures. The same holds true for most modern aluminum alloys, and for complex electronic components with their specific rare-earth elements, joined in configurations unthinkable under random geochemistry.

The total mass of steel produced by humanity exceeds several tens of billions of tons. A substantial share of this metal will, over time, oxidize, rust, and lose structural integrity — but even rusted-out metal leaves behind local anomalies in the concentration of iron and its accompanying alloying elements, often in proportions not found in natural ore deposits.

This brings us back to an idea first formulated in the fourth chapter, in connection with the Silurian Hypothesis: not the shape of an object, but an unusual combination of elements impossible under random natural processes — that is what is most likely to outlive us on a geologically meaningful timescale.

✦ A VOICE FROM THE STONE ✦

We hear how you speak of your footprint, and it strikes us as strange.

We never thought about what would remain after us, because we never separated ourselves from what surrounded us. Our trace and the forest's trace were the same trace. We left behind nothing the forest could not take back into itself.

You speak of steel, of plastic, of stone that does not soften in the rain. You take pride in this — in the fact that you will outlive yourselves in a form you yourselves would not recognize, when the time comes.

We would not take pride in that. But perhaps you don't need to. Perhaps it's enough for you to know that someone, someday, will find the trace — and understand that you were.

V. The Radioactive Signature: Invisible Ash

We already mentioned plutonium-239 as the official marker of the start of the Anthropocene. But humanity's radioactive footprint is more complex and more interesting than a single isotope.

Nuclear weapons tests and the operation of numerous nuclear reactors have left behind in the geological record not one but an entire complex of associated isotopic anomalies — including a uranium isotope ratio characteristic of nuclear detonations, cesium-137, strontium-90, and a number of other markers, short-lived on a geological scale but extraordinarily telling. In the regions of direct nuclear testing — such as the Nevada Test Site or Bikini Atoll — physical artifacts of a particular kind remain: trinitite, a glassy rock formed from sand melted by the blast, whose structure and isotopic composition have no equivalent in natural geological processes.

Most of these isotopes, as we have already noted, decay within tens of thousands of years — a negligible span on the scale of this entire book. But their global presence, recorded almost simultaneously on different continents in the mid-twentieth century, remains an exceptionally valuable example of exactly what a technological — rather than biological or geological — signal looks like: sharp, synchronous, global, and unexplainable by any natural process.

If the hypothetical ancient civilization we discussed in previous chapters had developed nuclear power, corresponding isotopes with longer half-lives could, in principle, have left a trace still detectable even after hundreds of millions of years. The absence of such a trace in the geological record known to us is not proof that no civilization existed, but it is certainly a serious argument against the existence of nuclear technology specifically, in its earthly history, if such a civilization existed at all.

✦✦✦

VI. The Chicken Bone as a Symbol of the Age

Perhaps the most unexpected example of humanity's technological footprint, almost comic in its down-to-earth banality, is the ordinary chicken.

In 2018, a group of biologists and geologists studying potential markers of the Anthropocene noted a remarkable fact: the modern broiler chicken — a breed developed by humans through intensive selective breeding for rapid industrial fattening — is a creature whose morphology differs so radically from its wild ancestors that its skeleton can rightly be considered a "technofossil," a fossil produced by technology rather than by natural selection.

The total number of chickens alive on Earth at any given moment today exceeds twenty-three billion — more than all other birds on the planet combined. Over sixty billion broilers are slaughtered every year. Their bones, ending up in landfills and dumps worldwide in colossal quantities, have a morphology radically different from any wild bird: disproportionately large breast muscles, a distorted skeletal structure, and a specific slaughter age recorded in the degree of tissue ossification.

The scientists who conducted this research concluded that if, a hundred million years from now, some future geologist stumbles upon a characteristic mass concentration of such bones within one thin geological layer, occurring worldwide at the same time, this would likely turn out to be one of the most recognizable, most unambiguous biological markers specifically of a technological, deliberately organized civilization — rather than merely an era with a lot of birds in it.

There is something deeply sobering in the thought that one of the primary pieces of physical evidence of our technological genius capable of surviving a hundred million years will turn out to be not a pyramid, not a skyscraper, not a spacecraft — but the deformed skeleton of a chicken.

"The broiler chicken may become the most vivid, tangible symbol of the Anthropocene — an epoch defined not by grandeur, but by mass production and industrial optimization."

— a paraphrased finding from a 2018 study, University of Leicester

VII. What Will Actually Outlive Us

Let's gather together everything we've examined in this chapter and try to give an honest answer to the question we started with: what will remain of humanity a hundred, two hundred, three hundred million years from now?

Not skyscrapers — they will be destroyed by erosion, and, if they're lucky enough to survive destruction, will be crumpled up and recycled by plate tectonics, as we learned in the second chapter. Not individual objects — hammers, tools, machines — the survival of a single object over such spans is statistically negligible, as we learned in the third chapter.

What will remain — with high probability — is a thin, but globally synchronous, geochemical layer: an anomalous concentration of rare metals and synthetic alloys; molecular traces of plastic fused into the sedimentary rock of deep ocean trenches; a sharp carbon isotope shift, similar in nature to the PETM we discussed in the fourth chapter — only caused not by volcanism, but by the burning of fossil fuels; and, perhaps, under favorable conditions, characteristic concentrations of the bones of domesticated species we have genetically modified.

In other words: exactly the same things we ourselves propose searching for in the quest for hypothetical predecessors. Without meaning to, we have run the perfect control experiment for our own methodology.

This is both sobering and reassuring. Sobering, because it shows just how modest, blurred, and easily overlooked our own technological footprint will turn out to be, despite all the grandeur we attribute to it today. Reassuring, because it confirms that the methodology proposed by Schmidt and Frank is genuinely aimed at the right kinds of traces — the kind that realistically survive — rather than at fantastical hopes of finding an intact city.

✦ ✦ ✦

VIII. A Comparison: If We Were Searching for Ourselves

Let's run this chapter's final thought experiment.

Imagine that humanity vanished today — for whatever reason, it doesn't matter which. Imagine further that, three hundred million years later, on a planet that has weathered several glacial cycles, several continental collisions, several major volcanic episodes, a new intelligent species emerges — say, descendants of the same crows or octopuses we discussed in the fifth chapter — and begins studying the geological record of its planet.

Will they find us?

In all likelihood — yes, but not in the form we would want to be found. Not as statues, books, works of art. They will find a strange, unnaturally sharp isotopic shift in the sedimentary rock of one particular thin layer. They will find inexplicable concentrations of metals not found in nature in such proportions. Perhaps, in exceptionally lucky, rare cases — a fragment of plastiglomerate at the site of a former coastline, miraculously preserved under protected, deeply buried sedimentary conditions.

They will very likely find not a single recognizable book, not a single recognizable statue, not one piece of evidence of our art, our philosophy, our emotional inner life — the very things we ourselves consider the true essence of human civilization. Of all of it, literally nothing discernible on a geological scale will remain.

This is the most honest, most unsettling lesson of this entire book, framed through the lens of ourselves: what we value most in ourselves is the least durable. What will survive geological epochs of us is a byproduct of our industry, not our soul.

✦ *A VOICE FROM THE STONE* ✦

If you find us — do not think you have found everything we were.

You will find the ash of what we made with our hands. You will not find the songs we sang to each other in the evenings, when the swamps grew quiet and the sky turned a darkness you will never see again.

You will not find how we feared. How we loved those we lost. How we looked at the stars and, probably, asked ourselves the same question — whether someone out there was looking back.

Remember this when you judge us by our ash. The ash is not us. The ash is only the small part of us that agreed to stay behind.

IX. From Matter to Perception

This chapter closes the first, more "technical" half of the book — the half concerned with the question of what can physically survive geological time, and how to search for it. We have traveled from grasping the scale of deep time, through understanding the mercilessness of the geological eraser, to examining specific artifacts, to the rigorous methodology of the Silurian Hypothesis, to the question of who might have any claim to intelligence before humanity, and finally, to an honest test of that methodology on ourselves.

But one more, far trickier question remains — not a physical one, but a philosophical one. Even if a trace exists. Even if it lies right in front of us, in an ice core, in sedimentary rock, in an isotope ratio — would we recognize it, if we saw it?

After all, as we learned in the fifth chapter, an alien intelligence might have left behind an alien trace — a technology built not on metal and concrete, but on biology and chemistry; not on individual genius, but on collective agreement. Is our perception flexible enough to recognize such a trace, even if it were staring us in the face?

It is precisely this question — the art of recognizing the alien, the most difficult and most philosophically charged problem in the entire book — to which the next chapter is devoted.

* * *

To be continued in Chapter Seven: "The Art of Recognizing the Alien"

Chapter Seven

The Art of Recognizing the Alien

Why We Search for a Mirror — and Why That's a Mistake

In the early sixteenth century, the European travelers who first journeyed up the Amazon into the continent's interior left behind rapturous accounts: densely populated riverbanks, rows of villages stretching, by some testimonies, for many kilometers, fields of cultivated crops, a complex web of settlements.

A century later, when the next wave of Europeans returned to the same places, they found something entirely different: impenetrable, wild jungle, almost uninhabited, without the faintest hint of civilization. These travelers — and, following them, centuries of European scholarship — drew what seemed like the logical conclusion: the first storytellers had exaggerated, had let their imaginations run away with them, and the Amazon had always been a pristine, human-untouched wilderness, a "green hell" where a complex civilization simply could never have arisen, given the poverty of tropical soils.

This version became the scientific consensus for a good four hundred years.

It turned out to be wrong. And exactly how it turned out to be wrong is one of the most instructive stories in the entire history of the human sciences — and it will serve as the foundation for this whole chapter.

✦ ✦ ✦

I. The Problem of the Observer

Before turning to the story of the Amazon, let's state the philosophical problem this chapter sets out to address.

In all the previous chapters, we searched for physical and chemical traces: artifacts, isotopes, geochemical anomalies. But the very idea of a "trace" assumes we're capable of recognizing it if we see it. What if that assumption is wrong? What if the key problem is not the absence of traces, but our inability to recognize them, even while looking straight at them?

Human perception is not a neutral, objective instrument for recording reality. It is evolutionarily tuned to recognize particular patterns — above all, the ones that mattered for our ancestors' survival: faces, predators, edible plants, signs of danger. Anything that doesn't fit our pre-set perceptual categories, our brains tend either to ignore, or, more insidiously, to automatically and unconsciously force into some category that already exists.

Applied to the subject of this book, this implies something unsettling: if an ancient technology were built on principles radically different from our own — not on metal and stone, but, as we speculated in the fifth chapter, on biology, ecosystem management, chemical communication — we could look directly at its traces and see nothing but "nature." Not because we're inattentive, but because the category "artificial, yet unlike our own technology" simply doesn't exist in our perceptual toolkit.

"We don't see the world as it is. We see the world as we are."

— Anaïs Nin

II. Terra Preta: The Soil That Turned Out to Be an Invention

Let's return to the Amazon.

Starting in the mid-twentieth century, and especially actively from the 1990s onward, archaeologists and soil scientists began paying close attention to a strange feature scattered across many parts of the Amazon basin: patches of unusually dark, extraordinarily fertile soil, in stark contrast to the poor, quickly exhausted lateritic soils typical of the tropics.

Locals had long called this soil terra preta — "black earth" — and prized it for its exceptional fertility, far surpassing that of the surrounding land. But the origin of this soil remained a matter of dispute for a long time: some researchers considered it the result of natural volcanic or river deposits, others simply a local geological anomaly.

Detailed analysis conducted in the twenty-first century has produced an unambiguous answer: terra preta is a soil artificially created by humans. It contains high concentrations of charcoal — so-called biochar — pottery shards, bones, and organic waste, deliberately worked into the soil by Amazonia's indigenous peoples over many centuries, and by some estimates, millennia, extending back to the pre-Columbian era.

This technology — in essence a form of deliberate, long-term soil-fertility engineering — made it possible to sustain intensive agriculture in territories that, under ordinary conditions, would be considered unsuitable for it, given how quickly tropical soils become exhausted. The extent of terra preta is estimated at anywhere from a few thousand to, by some bolder estimates, tens of thousands of square kilometers — evidence that we are looking not at some isolated experiment by a single settlement, but at a large-scale, widely distributed technological tradition.

For several centuries, European science looked at the richest, most fertile soil in the basin of the richest river on Earth — and saw in it nothing but a natural feature of the landscape. The true creator of this soil — deliberate human activity — was literally invisible to an observer who had already decided, in advance, that these jungles could not have hosted a developed civilization.

TERRA PRETA — ENGINEERING MISTAKEN FOR NATURE

Dark, fertile soil scattered across the Amazon basin, containing biochar, pottery, and organic remains

For centuries, thought to be a natural geological or riverine phenomenon

Modern analysis has confirmed: the result of deliberate, long-term indigenous agrotechnology

Estimated extent — from thousands to tens of thousands of square kilometers

The lesson: the artificial was mistaken for the natural across four centuries of scientific observation

✦✦✦

III. The Forest Once Thought Wild

The story of terra preta is only part of a far larger reassessment of our understanding of the pre-Columbian Amazon that has taken place over the last thirty years, thanks to a combination of archaeology, paleobotany, and — especially telling for our theme — the airborne laser-scanning technology known as LIDAR.

When researchers began using LIDAR to scan the Amazon jungle — a technology capable of "seeing" the terrain beneath dense vegetation cover — they discovered something that had been physically invisible to the human eye for centuries: the remains of geometrically regular earthworks, roads, canals, and residential platforms, stretching over enormous distances beneath what appeared to be untouched forest.

What's more, the very species composition of the Amazon forest turned out to be far less "wild" than previously assumed. Botanical research has shown that a substantial share of the species mix in the Amazon jungle is skewed toward so-called "domesticated" or "semi-domesticated" plant species — palms, fruit trees, plants useful to humans — occurring in these forests at concentrations statistically far exceeding what one would expect from a random, purely natural distribution of species.

In other words: what travelers and scientists took, for centuries, to be a virgin, untouched forest turned out to be, to a significant degree, the result of centuries of deliberate, yet extremely subtle, ecosystem management — management so organically woven into natural processes that it proved practically indistinguishable from them to an observer expecting to see either obvious untouched wilderness or the obvious signs of civilization in the familiar European sense — cities, monuments, writing.

The indigenous peoples of the Amazon did not conquer nature the way the builders of the Egyptian pyramids or the Roman aqueducts did. They wove their technology into nature so deeply that it became indistinguishable from nature itself — and that is exactly why it remained invisible to outside observers for four hundred years.

"We searched for ruins and saw jungle. The jungle was the ruin — simply a ruin that was still alive."

— an idea paraphrased from contemporary researchers of Amazonian archaeology

IV. Engineering Without Blueprints: Coral Reefs and Beaver Dams

The story of the Amazon shows us how easily a perfectly ordinary human technology can be mistaken for nature. But there is an equally instructive flip side to this problem: the capacity of non-human species to create structures so complex that the line between "natural phenomenon" and "engineered structure" starts to blur on its own, regardless of who's looking at it.

Take beaver dams. These are not simply piles of sticks — they are deliberately engineered hydraulic structures, capable of radically reshaping the landscape of entire river valleys, creating vast wetland ecosystems, and altering groundwater levels across dozens of hectares. Some beaver dams, documented in particular in Canada's Wood Buffalo National Park, exceed eight hundred meters in length — a structure that, had it been built by humans, would unquestionably be recognized as a serious feat of engineering.

Or coral reefs — colonial structures built by billions of individual organisms, each biologically simple in the extreme, whose collective activity over thousands of years shapes structures visible from space — the Great Barrier Reef stretches more than two thousand three hundred kilometers. Not a single coral polyp has the faintest notion of the reef's ultimate architecture. And yet the result of their combined, distributed activity is indistinguishable, in scale and complexity, from a conscious engineering design.

We call a beaver dam and a coral reef "nature" without a moment's hesitation — because we know who made them, and our cultural category for beavers and corals is "wild animals," not "engineers." But if we stumbled upon similar structures without knowing their origin, how would we classify them? This is not an idle play on words. It is precisely the problem science would face upon discovering a trace of the biological, ecosystem-based technology of the hypothetical ancient civilization we discussed in the fifth chapter.

✦ *A VOICE FROM THE STONE* ✦

We drew no border between ourselves and the forest.

You ask: where does your city end and the wilderness begin? We had no such question, because there was no such border. A tree we planted a hundred generations ago was no different, to us, from a tree that grew on its own. Both fed us. Both deserved our gratitude.

If you are looking for a wall — you will not find one. If you are looking for a seam between our design and natural growth — you will not find that either, because we never left a seam. We wove ourselves into the fabric so tightly that today, looking at it, you see only fabric.

Do not look for the shape of our presence. Look for a pattern where, in your view, there should only be chance.

V. Criteria for Recognizing Design

So how, then, is it even possible to tell the product of intelligent design apart from the result of purely natural — however astonishingly complex — processes? Is there an objective set of criteria, or are we doomed to endless, unresolvable arguments about exactly where the line falls?

The science of searching for extraterrestrial intelligence — the field known by the acronym SETI — has spent decades developing rigorous methodological principles for precisely this task: how to distinguish a signal produced by intelligence from one produced by natural astrophysical processes. And these principles, it turns out, apply far beyond the bounds of radio astronomy.

The first principle is narrowness of bandwidth. Natural processes — stars, pulsars, interstellar gas — typically produce signals broadly distributed across the frequency spectrum. An artificial, directed signal, by contrast, is usually extraordinarily narrow and specific — concentrated within a very particular band uncharacteristic of natural processes.

The second principle is excessive, unmotivated complexity. Natural processes, for all their capacity to produce astonishingly complex structures — think of terra preta, mistaken for geology, or beaver dams, or coral reefs — usually display a complexity explainable by relatively simple, repeating rules. Design, by contrast, often produces complexity that is excessive relative to a purely functional task — in other words, it spends resources on something beyond mere survival or reproduction.

The third principle is reproducibility according to an independent template. If the very same, extremely specific pattern turns up in locations far apart from and unconnected to one another, that is a far stronger argument for a single intelligent source or a single technological tradition than an isolated, one-off find.

This is exactly why the broad, widespread distribution of terra preta across the entire Amazon basin — rather than a single patch of dark soil — became such a weighty argument for deliberate technology rather than a chance geological coincidence.

CRITERIA FOR RECOGNIZING DESIGN (BY ANALOGY WITH SETI METHODOLOGY)

1. A narrow, specific signal — as against the broad, diffuse distribution typical of natural processes

2. Excessive complexity, not explainable by simple, repeating rules

3. Reproducibility of the same specific pattern across independent, widely separated locations

4. The absence, after careful checking, of an equally convincing alternative natural explanation

✦ ✦ ✦

VI. Survivorship Bias in the Search for Technology

There is one more, subtler psychological trap worth naming explicitly: survivorship bias, applied to the very concept of technology.

We modern humans are accustomed to a specific, rather narrow type of technology — the kind that developed within our own particular evolutionary and cultural history:

metalworking, the use of fire, building with hard, durable materials. This variety of technology has proven extraordinarily successful for our species — so successful that we unconsciously begin to treat it as the only possible, or at least the only "real," kind of technology there is.

But this is a classic survivorship bias: we see the single successful example that has reached us — our own technological civilization — and mistakenly treat it as a universal, inevitable template, instead of recognizing that it is merely one branch out of an enormous space of theoretically possible technological pathways, one that simply happened to be the one that survived to the moment when we can observe and analyze it.

If the Amazonian civilization, with its ecosystem-based, biological soil-fertility technology, had encountered colonization a few centuries earlier or later — if its written and monumental traditions (which, judging by recently discovered geoglyphs and settlement sites, do in fact seem to have existed) had survived in greater volume — perhaps we would today regard "soft," ecosystem-based engineering as a fully legitimate, equal type of advanced technology, rather than as an exotic exception that had to be rediscovered four centuries after it was nearly declared not to have existed at all.

What if the technological race we consider ourselves the winner of was never the only one? What if, alongside our own line — the line of metal, fire, and writing — other, equally legitimate, equally sophisticated technological traditions existed throughout the deep history of Earth, simply built on fundamentally different principles — and that is exactly why we stubbornly fail to recognize their traces?

"The history of science is the history of the constant expansion of the category of "the possible." Every generation considers the limits of its imagination final — and every succeeding generation proves it wrong."

— a common principle in the history of science

VII. What If We've Already Found It — and Didn't Recognize It?

Let's return briefly to the material of the third chapter — to OOPArts, the out-of-place artifacts. We examined them there from the standpoint of a physical object: a hammer, a cube, a sphere, a wedge. But what if the right question is a different one — not "have we found a strange object," but "have we found a strange pattern in living nature that we've written off as coincidence"?

Paleobotanists and paleoecologists occasionally run into anomalies that don't fit simple models of natural ecosystem evolution: unexpectedly sharp, synchronous shifts in flora composition on different continents at the same time; strange, statistically improbable coincidences in the paths by which different plant species were domesticated in regions far removed from one another; local, island-like zones of anomalous biodiversity unexplainable by climate or geology alone.

The overwhelming majority of such anomalies, on careful examination, turn out to have a perfectly convincing natural explanation — just as happened with most of the artifacts in the third chapter. Science, holding to the principle of Occam's razor, rightly prefers the simplest available explanation, one that requires no additional, unsupported entities.

But the story of the Amazon teaches us something important: sometimes the "simplest" explanation is not the same as the "correct" one. Sometimes the simplest explanation is simply the most familiar, the most intuitively comfortable one for an observer who lacks the perceptual category needed for a more complex, but true, alternative.

This is not a call to abandon scientific rigor and see design behind every natural anomaly — we already examined in detail, in the third chapter, where that kind of haste leads. It is a call for a subtler, more patient form of doubt: doubt not only about whether a mystery exists, but about our own capacity to recognize a mystery, even while looking it straight in the eye.

✦✦✦

VIII. The Humility of Perception

We come to the end of this chapter with a conclusion that may feel uncomfortable: there is no final, reliable algorithm that guarantees telling design apart from chance in every individual case. The SETI-like criteria we set out in Section V are a useful, but imperfect, tool. The story of the Amazon shows that even armed with the most rigorous scientific methods, we are capable of failing to notice a large-scale, widely distributed technology for centuries, if it doesn't fit our preconceived expectations.

And yet there is not only anxiety in this conclusion, but a certain freedom as well. If we acknowledge our own perceptual limitations as a fact, rather than as grounds for despair, we can work consciously and methodically to overcome them. This is, in fact, exactly how the mystery of terra preta was ultimately solved: not through a sudden flash of insight, but through the patient accumulation of data over decades, the application of new observational technologies — LIDAR among them — and scientists' willingness to revise deeply entrenched assumptions in the face of new facts.

Perhaps the best thing we can do, setting out to search for intelligence before humanity, is not to try to guess in advance what it ought to look like, but instead to search systematically and patiently for pattern — narrowness of signal, excessive complexity, repeatability of pattern — regardless of how "natural" the context in which that pattern turns up might seem at first glance.

"The great obstacle to discovery is not ignorance — it is the illusion of knowledge."

— attributed to Daniel Boorstin

In the next chapter, we will rise to the highest, most abstract level in this entire book — from the geology and biology of one particular planet to a question of cosmic scale. If intelligence really is capable of arising on Earth repeatedly, independently, in the most varied biological forms, why does the universe look so empty? Why, after decades of deliberate searching, gazing into the night sky, have we still not encountered a single signal from outside?

This question is known to science as the Fermi Paradox — and, as we will see, it turns out to be unexpectedly closely intertwined with everything we've discussed in this chapter.

* * *

To be continued in Chapter Eight: "The Cosmic Filter and Earth's Silence"

Chapter Eight

The Cosmic Filter and Earth's Silence

Why the Universe Doesn't Answer — and What Earth Itself Knows About It

In the summer of 1950, several physicists — among them Enrico Fermi, one of the creators of the first nuclear reactor — were having lunch in the cafeteria at Los Alamos National Laboratory, engaged in a casual conversation about recent UFO reports and the possibility of interstellar travel.

The conversation went on as such conversations do, until Fermi suddenly asked a question that has borne his name ever since, and remains one of the most discussed in the history of science. By various accounts, the question went something like this:

"So where is everybody?"

The logic behind this short question is deceptively simple. Our galaxy contains something on the order of one hundred to four hundred billion stars, and has existed for roughly thirteen billion years. Even under modest, conservative assumptions about the speed of interstellar travel — say, a tenth of the speed of light — a civilization that had mastered space travel could, in theory, colonize the entire galaxy within a mere few million years: a negligible span on a cosmic scale. If intelligent life arises in the universe with even a modestly noticeable probability, the galaxy should long since have been teeming with civilizations that left behind countless observable traces of their activity.

But the sky is silent.

This silence is the Fermi Paradox: a sharp, unexplained discrepancy between what logic tells us to expect and what we actually observe.

✦ ✦ ✦

I. An Equation That Doesn't Add Up

Back in 1961, the astronomer Frank Drake proposed a formula, later named after him — an attempt to break down the complex question of how many technological civilizations exist in the galaxy into a set of simpler variables, at least roughly estimable: the rate of star formation, the fraction of stars with planets, the fraction of planets suitable for life, the probability of life arising on such a planet, the probability of intelligence developing, the probability of developing technology capable of interstellar communication, and, finally, the average lifespan of such a technological civilization.

The Drake Equation is not a rigorous mathematical tool in the usual sense. It is, rather, an organizing framework for having an honest conversation about our own ignorance: for decades, most of the variables that go into it remained essentially inaccessible to empirical estimation, and the equation's final result could come out as anything at all — from "we are alone in the entire observable universe" to "the galaxy must be teeming with civilizations" — simply depending on how optimistic or pessimistic the values a researcher plugged into the formula happened to be.

Over the last thirty years, the situation has changed considerably: thanks to space telescopes, above all the Kepler mission, we now know that planets around other stars are not a rarity, but rather the rule. The overwhelming majority of stars in the galaxy, by current estimates, have at least one planet, and planets within the so-called "habitable zone" — at a distance from their star that allows for the existence of liquid water — occur frequently enough that we can confidently speak of billions of potentially life-supporting worlds in our galaxy alone.

This, one would think, should have made the Fermi Paradox even sharper: the more we learn about how common potentially habitable planets are, the harder it becomes to explain the absolute, universal silence of the cosmos.

THE SCALE OF THE PROBLEM

100–400 billion stars in the Milky Way galaxy alone

Age of the galaxy: roughly 13 billion years

By current estimates, most stars have planetary systems

Even at 10% of the speed of light, colonizing the entire galaxy would take on the order of a few million years

The observed result of decades of SETI programs: not a single confirmed technological signal

II. The Great Filter: Where Is It Hiding?

In 1996, the economist and futurist Robin Hanson proposed a concept that has since become one of the most influential ways of making sense of the Fermi Paradox — the idea of the "Great Filter."

Hanson's logic runs as follows: if the path from a lifeless planet to a technological civilization capable of colonizing the galaxy consists of a series of sequential stages — the emergence of simple life, the emergence of complex multicellular life, the emergence of intelligence, the emergence of technology, the survival of a technological civilization for long enough to achieve interstellar expansion — then somewhere along this path there must exist one or several stages that are statistically extraordinarily improbable. So improbable that the overwhelming majority of potential civilizations get "stuck" at precisely this stage and never make it through successfully.

This critical, nearly insurmountable stage is what came to be called the Great Filter.

The key, unsettling question is: where exactly is this filter located, relative to us? If it lies behind us — that is, if, say, the mere fact of life's origin, or the emergence of complex multicellular life, is statistically extraordinarily rare in the universe — then humanity has good news: we have already passed the hardest test, and our path to the stars is, in principle, open.

But if the filter lies ahead of us — if, say, any civilization reaching our level of technological development has a high probability of destroying itself before it manages to reach space — then we, in all likelihood, face the same fate that befell all, or nearly all, other civilizations before us.

"The silence of the cosmos may be either the most reassuring or the most terrifying news humanity has ever received — depending on exactly where the filter is located."

— a paraphrased formulation of Robin Hanson's concept

✦✦✦

III. Earth as a Laboratory for the Filter

And here is where a genuinely unexpected turn begins — the very reason, in fact, this book has made its way to an eighth chapter: Earth, the one planet in the entire known universe whose history we can study down to the finest geological detail, is itself a miniature laboratory for testing the Great Filter hypothesis.

Let's recall the planet's history, which we have already partly examined in previous chapters. Life arose on Earth remarkably quickly by geological standards — apparently within a few hundred million years of the planet becoming capable of supporting liquid water at all. Some researchers, including Adam Frank, coauthor of the Silurian Hypothesis, interpret this observation as an argument that the mere emergence of the simplest life is probably not an exceptionally rare event — at least, not rare enough to serve as a full-fledged Great Filter.

But the path from the simplest life to complex, multicellular life took not hundreds of millions, but billions of years — nearly three-quarters of the planet's entire history. This is a far more serious candidate for the role of the filter: perhaps the transition from simple single-celled forms to complex, specialized multicellular organisms is a statistically extremely improbable event, requiring an exceptionally fortunate confluence of biochemical circumstances.

And then — and this is what directly connects this chapter to the entire preceding part of the book — over the following five-hundred-odd million years, starting with the Cambrian Explosion, the planet endured at least five mass extinctions of planetary scale, each of which destroyed a significant, and at times overwhelming, share of all complex biological diversity.

Each such extinction is, in essence, a reset: a zeroing-out of whatever evolutionary progress toward intelligence and complexity had accumulated up to that point. And if — purely hypothetically, recalling the fifth chapter of this book — at any of these stages, before a mass extinction, a lineage genuinely existed that was moving toward something resembling intelligence, a planetary-scale catastrophe could very well have cut that path short before it managed to culminate in anything technological.

THE FIVE GREAT EXTINCTIONS AS POTENTIAL EVOLUTIONARY "RESETS"

Ordovician (~444 million years ago) — up to 85% of species

Devonian (~372 million years ago) — a large share of marine fauna

Permian (~252 million years ago) — up to 96% of marine species, the largest extinction in history

Triassic (~201 million years ago) — cleared the way for dinosaur dominance

Cretaceous–Paleogene (~66 million years ago) — the death of non-avian dinosaurs, opening the way for mammals

Each event: a potential breaking point for any line of complexity developing up to that moment

IV. Frank's Argument: Earth Has Had Several Tries

Adam Frank, whose work on the Silurian Hypothesis we examined in detail in the fourth chapter, developed this idea into a separate, broader line of research devoted to the connection between Earth's history and the statistics of technological civilizations arising in the universe as a whole.

His argument, put simply, runs as follows: if we want to estimate the probability that a technological civilization has arisen somewhere else in the universe, we have at least one planet with a fully documented biological history, studied down to the finest detail, spanning several billion years — Earth itself. And this history is not a single, linear march toward intelligence, but a series of many independent evolutionary "attempts," separated by catastrophes.

After each mass extinction, life on the planet essentially restarted the evolutionary race from a blank slate, with a new set of dominant lineages inheriting the vacated ecological niches. After the Permian extinction, the previously dominant forms were replaced by the archosaurs, including the lineage that later gave rise to the dinosaurs. After the extinction of the dinosaurs, the vacated niches were filled by mammals — the very lineage that, millions of years later, produced us.

If we look at Earth's history through precisely this lens — as a series of perhaps half a dozen separate, largely independent evolutionary experiments, divided by global catastrophes — the statistics become genuinely interesting. Of all these independent "attempts," only one — the most recent, our own — appears to have produced a technological civilization capable of asking questions like these.

This can be interpreted in two opposite ways. The optimistic reading: since at least one of several independent attempts successfully reached the level of intelligence and technology, intelligence is not an exceptionally rare, near-impossible event, and somewhere else in the universe a similar experiment has probably succeeded too. The pessimistic reading: since only one of several independent attempts succeeded, and that the most recent one, the probability of success for any single attempt is extremely low — and success itself may not be a guarantee of technology's long-term survival, but merely a rare stroke of luck whose ultimate fate is still very much undecided.

✦ ✦ ✦

V. The Self-Destruction Hypothesis: Technology as a Weapon Turned on Itself

Here we arrive at the most disquieting part of this chapter — the possibility that the Great Filter lies not behind us, but ahead.

There exists an entire class of hypotheses united by a common idea: any civilization reaching a certain level of technological development automatically gains access to means capable of bringing about its own destruction — nuclear weapons, uncontrolled climate change, biotechnological risks, potentially dangerous uncontrolled artificial intelligence. If the probability of self-destruction at this technological stage is sufficiently high — say, if the overwhelming majority of civilizations fail to survive the first few centuries after mastering nuclear energy — this alone would explain the observed silence of the cosmos: civilizations arise, reach a technological peak, and then vanish rapidly, almost instantaneously by cosmic standards, without managing to spread noticeable technological signals beyond their own star system.

Here the Silurian Hypothesis from the fourth chapter converges with the Fermi Paradox in the most unexpected way. If technological civilizations tend to burn out quickly, brightly, and unnoticed — if their technological phase, like our own industrial age, lasts not millions of years but a mere few centuries or millennia — then the very brevity of this phase becomes an additional explanation for why we find no trace of them, either in the cosmos or, returning to this book's central theme, in the depths of Earth's geological record.

Perhaps it's not only that the geological eraser described in the second chapter wipes away the traces of ancient civilizations. Perhaps those civilizations themselves, if they existed, simply never lasted long enough to leave behind a trace worth erasing in the first place.

✦ *A VOICE FROM THE STONE* ✦

We did not burn out quickly. We faded slowly — so slowly that not one of us noticed the exact day it happened.

There was no last battle. There was no last cry. There was simply a little less rain, year after year, a little less fruit on the trees, a few more children who did not live to the age when the songs of the ancestors are taught.

We did not win or lose against anything in particular. We simply found ourselves part of a world that was changing faster than we ourselves were changing.

If this is what you mean by your "filter" — it does not look like a wall. It looks like a door slowly swinging shut, one that the children, one day, simply stop making it through in time.

VI. Silence as Data, Not as Emptiness

Many people share a common intuitive mistake in how they perceive the silence of the cosmos: treating it as an empty space, as an absence of information. In reality, from the standpoint of rigorous statistics, the absence of a signal is itself data — and quite informative data, if analyzed correctly.

Decades of deliberate searching — including large-scale radio observation programs, the analysis of telescope data for artificial technological signatures, the search for the telltale signs of large-scale engineering structures such as hypothetical Dyson spheres — have turned up not a single convincing, reproducible piece of evidence of technological activity beyond Earth. This is a robust, statistically significant null result.

A null result in science is not a failed experiment. It is valuable information that places strict constraints on the possible models of reality. Applied to the Fermi Paradox, this means: whatever the true rate at which technological civilizations arise in the universe, it evidently is not high enough — or not long-lived enough — to produce signals detectable at our current, still fairly limited, level of observational capability.

This brings us back to the central idea of this book in its sharpest form: perhaps intelligence is not rare in any absolute sense. Perhaps what is rare is intelligence's durability. Perhaps flashes of complexity like our own civilization ignite across the universe — and on Earth, if the logic of this book holds — far more often than is commonly believed, but burn out so quickly by geological and cosmic standards that the chance of catching one of them at its technological peak is statistically negligible.

"We may not be alone across time, even while remaining alone in the moment of observation. The universe may be full of ashes where we are searching for campfires."

— an idea paraphrased from contemporary astrobiological literature

✦✦✦

VII. The Drake Equation, Rewritten for a Single Planet

What if we applied the logic of the Drake Equation not to the entire galaxy, but to a single planet — to Earth itself — treating its deep geological history as a series of separate, independent experiments, the way Adam Frank proposes?

Such an exercise will, of course, remain largely speculative — we simply don't have enough data to plug in any reliably solid numbers. But the very structure of the question is extraordinarily instructive. How many times over the last five hundred million years did conditions on the planet allow some lineage of organisms to approach the threshold beyond which the path to intelligence and technology begins? How many times was that path cut short — by catastrophe, by competition, by the plain absence of sufficient evolutionary pressure — before reaching the finish line?

We already examined several potential candidates in the fifth chapter: the giant arthropods of the Carboniferous, with their biological ceiling on brain growth; the early tetrapods, who never encountered sufficient pressure toward complex social behavior; the collective insects, with their fundamentally different, distributed model of problem-solving. Not one of them, as far as we can tell from the available data, crossed the finish line. But the mere existence of these candidates is evidence that the preconditions for such experiments arose on Earth more than once, not just a single time.

Perhaps the right conclusion to draw from Earth's history is not "intelligence is extraordinarily rare," but "intelligence tries itself out repeatedly, but extraordinarily rarely makes it all the way." This is a fundamentally different, more encouraging — though no less unsettling — formulation of the very same phenomenon.

VIII. What the Silence Says About Us

We come to the end of this chapter with a question that inevitably brings us back from the cosmos and the deep geological past to the present moment — to ourselves, reading this book here and now.

If the Great Filter really does lie ahead of us, and not behind — if the statistics of Earth's geological history, the statistics of many broken-off evolutionary attempts, and the statistics of the cosmos's staggering silence add up to a single, disquieting picture — then the most urgent question is not "did an ancient civilization exist before us," but "will we ourselves make it through the filter that may have stopped every single one before us"?

This shift in focus — from the archaeology of the past to the responsibility of the present — closes out the book's eighth chapter deliberately, not by accident. Because the entire logic we've built up across the preceding seven chapters — the mercilessness of the geological eraser, the fragility of any physical trace, the narrowness of our perception when it comes to recognizing an alien intelligence, the statistics of multiple broken-off evolutionary attempts — converges on a single point: intelligence and technology are not a guaranteed, safe, inevitable outcome of the development of complexity. They are a rare, vulnerable, possibly self-destructing flash against the backdrop of the far more durable silence of the planet and the cosmos.

And if this is indeed so, the greatest value of this book lies not at all in answering the question of whether a civilization existed on Earth before us. It lies in this question forcing us to look more honestly at ourselves — at the one attempt of intelligence known to us that has not yet come to an end.

"Perhaps we are asking the universe the wrong question. Not "where is everybody?" but "how much time do we have before we, too, fall silent?""

— a closing thought for this chapter

In the next, penultimate chapter of the book, we will conduct one final, most personal thought experiment: imagining a far-future archaeologist studying the layer left by humanity three hundred million years after our disappearance. We will draw together everything we've learned in the previous chapters — geology, chemistry, biology, the philosophy of perception, the statistics of the Great Filter — to picture, in detail, exactly what this last, thinnest sliver of our own existence will look like, inscribed within the infinitely longer history of a silent planet.

* * *

To be continued in Chapter Nine: "Archaeology of the Year 300 Million"

Chapter Nine

Archaeology of the Year 300 Million

A Thought Experiment That Changes Everything

Let three hundred million years pass.

It doesn't matter who exactly does the digging — some new lineage of intelligent descendants of Earth's creatures, having survived catastrophe by chance, or visitors from another star, arriving at our planet after it has cooled and reheated once again. Let's call this being simply the Investigator — faceless, speciesless, nameless, because for our purposes what matters is not its nature, but what it encounters.

The Investigator knows nothing about us. It has no history textbooks, no photographs, not a single word in any human language. All it has is what this entire book has been devoted to: geology, chemistry, patience, and method.

We have already gathered, piece by piece, from the eight preceding chapters, everything needed to picture honestly, down to the details, exactly what this Investigator will find. It's time to assemble those pieces into a single picture.

✦✦✦

I. A Section That Looks Like Nothing Special

The Investigator's expedition does not begin with a sensation. It begins with boredom — with the routine, scheduled geological mapping of some region of a planet that, over the intervening three hundred million years, has, by our current projections, undergone at least a partial breakup of today's continents, a collision between Africa and Europe, the closing of the Mediterranean Sea, the formation of new mountain ranges, and, quite possibly, the beginning of the assembly of a new supercontinent, one geologists have already given a provisional name today — Pangaea Ultima.

Where today lies, say, the coast of modern Florida, or the delta of one of the great Asian rivers — terrain that, over the intervening epochs, has repeatedly sunk below sea level and risen again — the Investigator, drilling yet another exploratory borehole through a body of sedimentary rock, stumbles upon something unusual.

Not gold. Not a skeleton. Not the gleam of metal. A thin layer of rock — literally a few centimeters — different from everything lying above and below it: a layer in which the concentration of certain elements deviates sharply, anomalously, statistically, from the background level.

This is exactly what we, armed with the logic of the fourth and sixth chapters of this book, would have predicted in advance: not a city, not a building, not a mechanism. The thinnest strip of geochemical anomaly, inscribed within kilometers of perfectly ordinary, entirely unremarkable sandstone and clay.

II. What the Drill Will Show

The first laboratory analysis of the sample the Investigator has extracted registers a sharp shift in carbon isotope ratios, localized entirely within this narrow layer — the very "negative excursion" we discussed in detail in the fourth chapter in connection with the PETM. The rock in this layer is statistically significantly depleted in the heavy isotope carbon-13, compared with the layers immediately above and below.

The second analysis — a mineralogical one — finds, in that same thin layer, an anomalous, elevated concentration of several metals: iron, aluminum, rare elements such as neodymium and lanthanum, in proportions statistically uncharacteristic of purely natural geochemical processes in this region. In certain parts of the layer, microscopic, extraordinarily durable particles of synthetic polymer compounds turn up — degraded, but still chemically recognizable plastic, preserved in a thin, protected pocket of anaerobic, oxygen-free sedimentary conditions, the kind we discussed in the sixth chapter.

The third, most sensitive analysis — an isotopic analysis of trace radioactive elements — registers residual, almost indistinguishable, but statistically still detectable traces of decay in isotopes with exceptionally long half-lives: small but orderly anomalies in the ratio of uranium to its decay products, pointing, evidently, to localized zones of once extraordinarily intense, artificially organized nuclear activity — an echo of the very reactors and test sites discussed in the sixth chapter, stretched across three hundred million years of decay, but still, in traces, discernible to a sufficiently precise instrument.

Not one of these three signals, on its own, would be enough for a sensational conclusion. But all three, found within the same narrow geological layer, synchronously, at several sites scattered across the planet — that is precisely the combination Schmidt and Frank, in the fourth chapter, would call serious grounds for further, most careful investigation.

WHAT THE DRILL WILL FIND THREE HUNDRED MILLION YEARS FROM NOW

A sharp carbon isotope excursion — resembling the PETM signature, but of a different origin

An anomalous concentration of metals and rare-earth elements in uncharacteristic proportions

Microscopic, durable molecular traces of synthetic polymers

Residual decay-isotope anomalies pointing to once-intense nuclear activity

Thickness of the entire anomalous layer: likely a few centimeters, set against kilometers of surrounding sediment

✦ ✦ ✦

III. Cities That Became Geology

What will happen, over three hundred million years, to the cities themselves — to New York, Tokyo, Moscow, to all the infrastructure we today consider the pinnacle of our species' engineering genius?

Drawing on everything we examined in the second chapter, the answer is known in advance: practically nothing of them will survive in recognizable form. Buildings made of steel and concrete will begin to collapse within a few decades, a few centuries at most, once maintenance stops. Water will work its way into cracks, rust will eat through rebar, plant roots and freeze-thaw cycles will split concrete slabs apart. Within a few millennia, what stands where a modern megacity once stood will most likely be nothing but a chaotic field of rubble, gradually buried under soil and vegetation.

Within a few tens of thousands of years — unless the city is buried beneath new sedimentary layers early enough and deeply enough to escape prolonged erosion — even the memory of its geometry, its streets, its buildings, will vanish completely. All that will remain is a chaotic, statistically anomalous concentration of construction debris: chunks of concrete, rusted metal structures, fragments of glass and ceramic, mixed in among ordinary sedimentary deposits.

But if — which is far more likely for low-lying coastal cities, some of which will probably already be underwater by then due to rising sea levels — the city's rubble is buried quickly enough beneath a layer of silt and sediment, we get a chance at something more interesting: a kind of artificial conglomerate, in which fragments of brick, concrete, glass, ceramic, and metal are all mixed together — a distinctive anthropogenic breccia, the geological equivalent of the plastiglomerate discussed in the sixth chapter, only on a far greater scale.

The Investigator, stumbling upon such a conglomerate, will most likely not recognize it as a "city" in the sense familiar to us. It will see only an unusually dense, statistically anomalous cluster of heterogeneous materials, clearly not of natural origin — precisely what we discussed in the seventh chapter in connection with the criteria for recognizing design: narrowness, anomalous density, the reproducibility of similar clusters across different, widely separated parts of the planet.

"Cities do not die in a single instant. They crumble into a crowd of debris, and then into a crowd of minerals, indistinguishable from any other rubble on the planet, unless you know exactly what to look for."

— an idea paraphrased from contemporary literature on the geology of the Anthropocene

IV. The Chicken, the Glass Trace, and the Excavator

Among the Investigator's most striking, though at first glance most trivial, finds may be something we already discussed in the sixth chapter: a concentration of bones with a statistically anomalous morphology, unlike anything else.

In the sedimentary layer corresponding to our era, scattered across numerous far-flung continents, the Investigator, if lucky, will find mass burials of the bones of one and the same remarkable bird — with disproportionately developed breast musculature, a clearly defined, statistically repeating slaughter age, occurring in numbers utterly incompatible with any model of a species' natural, non-intelligence-managed spread. This, as we noted earlier, is one of the most unambiguous biological markers of organized, industrial activity specifically.

In certain, especially fortunate cases — where large-scale fires once occurred, or, more likely, where nuclear weapons tests once took place — the Investigator may stumble upon localized patches of hardened glassy material similar to trinitite, which we discussed in the sixth chapter: a rock whose isotopic composition and structure unambiguously point to a brief but extremely intense thermal event, with no equivalent among the known natural processes of the region.

And in the rarest, almost improbable stroke of luck, the Investigator might stumble upon something like a modern-day version of the Coso Artifact from the third chapter: a single, exceptionally well-preserved fragment of a mechanism, preserved under uniquely favorable anaerobic conditions — perhaps an engine part, perhaps a fragment of a circuit board — encased in a mineral crust formed over millions of years of burial. And here we inevitably return to the central methodological lesson of the third chapter: such a singular find, however sensational it may seem, will require the most rigorous, most scrupulous verification before it can be accepted as genuine evidence, rather than a chance coincidence or a misinterpretation.

✦✦✦

V. Arguments in the Laboratory of the Future

Let's imagine that all these finds — the isotopic excursion, the metal anomalies, the concentrations of chicken bones, the isolated fragments of glassy rock — have been gathered together and placed before the Investigator's scientific community.

Will unanimous declarations of an ancient technological civilization's existence follow immediately? Most likely not — and, as we examined in detail in the third and seventh chapters, this would be the correct, healthy scientific response, not a sign of rigidity or narrow-mindedness.

One group of skeptically minded researchers will quite likely propose alternative, purely natural explanations for each individual signal: the carbon isotope excursion will be attributed to yet another, not-fully-understood episode of volcanic activity or methane release — after all, as we know from the fourth chapter, such natural events really have occurred more than once in the planet's history. The metal anomalies will be explained as a local, unusual type of ore deposit. The concentrations of bird bones — perhaps as a colony of some previously unknown species, wiped out by a local catastrophe.

A second group, more inclined to see something more in these finds, will insist: there are too many coincidences, too much synchrony among different types of signals, too broad and global a distribution of one and the same narrow, specific layer, for it to be explained solely by random, local natural processes.

This argument will, in all likelihood, drag on for more than a single year, more than a single decade of the Investigator's own internal chronology — just as our own scientific debates over the origin and nature of the Carboniferous artifacts, over the origin of the Klerksdorp spheres, over the correct interpretation of the Silurian Hypothesis, have continued for decades without yet reaching a final, universally accepted resolution.

✦ *A VOICE FROM THE STONE* ✦

Perhaps someday, someone will argue about us the way you are arguing about us now.

They will find our ash and split into two camps: some will say — this is just stone, just chance, just a trick of nature. Others will say — no, there are too many coincidences here, there is a hand felt here, there is a thought felt here.

And both camps will be partly right. Because we were — and, at the same time, we ceased to be anything more than stone. Time makes both answers true at once.

Do not hurry to a conclusion, when your own turn comes to be found. Let them argue for a long time. The argument is the only memory we will have left.

VI. Reconstructing the Unthinkable

Let's suppose — allowing ourselves a moment of optimism — that the Investigator's scientific community eventually reaches a consensus: yes, what lies before us are the traces of a once-existing, once-technologically-advanced civilization. What, exactly, will they be able to reconstruct about us from this scant, scattered, geologically mangled data?

They will probably establish, with fair confidence, that this civilization had industry, used fossil fuels, possessed metallurgy, and, at least in the late stage of its existence, nuclear technology. They might be able to estimate the approximate duration of this technological phase — from the thickness and internal structure of the anomalous layer — and arrive at a conclusion strikingly similar to the one we ourselves formulated in the sixth chapter: the technological phase was geologically extraordinarily brief, a flash rather than a long epoch.

But here is what they will, in all likelihood, never learn — and this is the most piercing, most philosophically charged conclusion of the entire book: they will not learn how we spoke to one another. They will not learn our names, our languages, our religions and philosophies. They will not learn that we wrote poems, composed music, fell in love, feared the dark, mourned our dead, laughed at jokes, looked up at the stars, and made up stories about what was up there.

We already arrived at this conclusion at the end of the sixth chapter, but now, seeing it through the eyes of the Investigator, armed with the full methodology of the preceding eight chapters, it lands even more sharply: the technological trace outlives us. The inner, human essence does not.

And if this is true of us — the most thoroughly documented, most studied, most self-assured civilization ever to have existed on this planet — what are the odds that we, studying the hypothetical predecessors from the fourth and fifth chapters of this book, could ever learn anything at all about what they were like on the inside, even if we found irrefutable proof of their technological existence?

"Geology preserves what we did. It does not preserve who we were."

— a closing thought for this chapter

✦ ✦ ✦

VII. A Lesson for Ourselves

This thought experiment, for all its speculative nature, serves a far more practical purpose than simply being a nice metaphor. It amounts, in essence, to a complete, detailed model of exactly what a real search for traces of an ancient Earth civilization should look like — and exactly how it should be interpreted — should we ever seriously undertake it ourselves.

We now know what's not worth searching for: cities, skeletons, recognizable mechanisms — the very mistake made by the OOPArts enthusiasts of the third chapter, who spent decades trying to pass off concretions and meteorites as proof of intelligent design.

We know what is worth searching for: narrow, synchronous, globally reproducing geochemical anomalies — isotopic excursions, unusual metal concentrations, traces of durable synthetic compounds, residual radioactivity.

We know that even upon discovering such a signal, the correct scientific response is not sensationalism, but a long, patient, skeptical argument, in which the simple natural explanation must be exhausted before any alternative is seriously entertained — exactly as would happen in the Investigator's laboratory.

And finally, we know — perhaps this is the most important, if also the saddest lesson — that even in success, even upon discovering an irrefutable technological trace, we will most likely never learn who its makers were on the inside. We will learn only that they existed, that they were trying to build something — and that, sooner or later, like us, they fell silent.

VIII. A Letter in Stone

There is a peculiar, almost unbearable beauty in the thought that this entire book, all nine of its chapters, all its patient, disciplined curiosity, is, in the end, nothing more than an attempt to write one very long, very slow letter.

A letter addressed not to the reader of this book, but to that very hypothetical Investigator, who may never appear at all — and if it does appear, it will be three hundred million years from now, when nothing of us ourselves will remain but a thin, anomalous layer within a body of nameless sedimentary rock.

We do not know whether anyone will ever read this letter. We are not even entirely sure we know how to write it so that it will be read correctly — after all, as we found in the seventh chapter, recognizing an alien intelligence is a task of nearly impossible difficulty even when applied to our own biological species, let alone to someone separated from us by an age as long as geological eternity.

And yet we write it anyway — with every factory, every power plant, every weapons test, every ton of plastic that ends up in the ocean. We write it not because we chose to, consciously, but because we cannot do otherwise: technological activity leaves a trace as inevitably as breathing leaves a trace of carbon dioxide in the air.

✦ *A VOICE FROM THE STONE* ✦

We, too, were writing a letter, without knowing it.

Every cleared patch of forest, every bone charred on a fire, every handful of ash left where grass once grew — these were the letters of our message, though we never thought of ourselves as scribes.

We do not know who will read our letter. Perhaps you are already reading it — you simply haven't realized yet that you're holding it in your hands.

If that is so — know one thing: we were not monsters, and we were not gods. We were the same as what you are right now. Beings who lived, who feared disappearing, and went on living all the same.

In the final, tenth chapter of this book, we will at last step back from specific hypotheses, specific methods, specific thought experiments — and allow ourselves the very thing this whole book was, in essence, written for: to pause, and simply feel the scale of everything we have discussed across these nine chapters.

Because, as we will see in the finale, the meaning of this book never lay solely in the search for evidence. It lay in a lesson of humility — the very lesson the planet patiently teaches anyone who gazes long enough into the depths of its silence.

* * *

To be continued in Chapter Ten: "A Lesson in Humility Before Eternity"

Chapter Ten

A Lesson in Humility Before Eternity

On Why We Need a Book With No Final Answer

Take a pencil in your hand.

The very one we spoke of at the beginning of this book — twenty centimeters long, wooden, yellow, with a pink eraser at the end. Since then, we have come a long way: from the geological eraser that wipes out entire worlds, through strange artifacts, isotopic excursions, and giant dragonflies, to a silent cosmos and an imagined Investigator who might, one day, find our own trace.

It's time to look at this pencil again — but with different eyes now.

Because over nine chapters, we were not so much searching for an answer as learning how to look properly at the question.

✦ ✦ ✦

I. What We Know for Certain

Before allowing ourselves the philosophical conclusion this entire book has been leading toward, we need, one last time, with complete honesty, to separate what has been established from what remains hypothesis.

We know for certain: geologic time is far deeper and stranger than human intuition can register — and this is not a metaphor, but a measurable fact arising from the physics of the planet itself. We know the planet possesses extraordinarily powerful mechanisms for destroying its own history — plate tectonics, erosion, metamorphism, volcanism, glaciation — and that these mechanisms make the absence of physical artifacts from antiquity a statistically expected phenomenon, not proof that there is nothing to find.

We know that the overwhelming majority of "sensational" finds called OOPArts, under rigorous scrutiny, turn out to have perfectly mundane explanations — but we also know that at least a few cases, like the Aiud Wedge, remain unresolved today not because they prove something extraordinary, but because the available data is simply insufficient for a definitive conclusion in either direction.

We know that serious scientists — Gavin Schmidt and Adam Frank — formulated a rigorous, testable methodology for searching for traces of a possible ancient technological civilization, based not on artifacts, but on geochemical anomalies like the PETM signature. We know this methodology has so far produced not a single convincing positive result — but we also know the geological record has been studied too unevenly for this null result to be considered final.

We know that, deep in Earth's past, biological lineages existed — giant arthropods, early tetrapods, collective insects — that possessed one precondition or another for the development of complex behavior, but not one of them, as far as we know, crossed the threshold separating complexity from technology.

And finally, we know — this is perhaps the single most solidly established fact in the entire book — that human perception systematically underestimates the likelihood of non-human, non-human-like intelligence and technology, as shown by the history of the European "discovery" of the Amazon, which remained invisible engineering wearing the mask of wilderness for four centuries.

SUMMARY: WHAT IS ESTABLISHED, WHAT REMAINS OPEN

Established: the depth of geologic time, the mercilessness of geological processes of destruction

Established: a methodology for searching for chemical, rather than physical, traces of a possible civilization

Established: nature's capacity to produce forms mistaken for artifacts — and technology's capacity to disguise itself as nature

Open: whether a technological civilization ever existed before humanity

Open: whether the Great Filter lies behind us or still ahead

II. Humility Is Not the Same as Despair

There is a temptation to read this entire book as a chronicle of defeat: we searched, and searched, and searched — and found nothing conclusive. No confirmed civilization, no disproven hypothesis, no clear answer to the question of whether we have been alone on this planet across its whole long history.

But that is a misreading. Because the purpose of this book was never to provide a final answer. It was to show that the question itself is legitimate, serious, and deserving of the most rigorous scientific attention — not mocking condescension.

The humility this chapter's title speaks of is not a surrender to the unknown. It is a particular kind of intellectual maturity: the ability to hold two seemingly contradictory things in mind at once. First: we will most likely never get a final answer to the question of ancient civilizations, because the geological record is too incomplete and our perception too limited. Second: this is not a reason to stop asking, but a reason to ask more precisely, more patiently, more honestly.

Real science, in the end, rarely gives final answers. It gives ever more precise questions. And if this book has taught the reader anything, it is not a set of facts about the Carboniferous or the Silurian Hypothesis, but the capacity to ask more honest questions about one's own place in time.

"Science is not a collection of truths. It is a method for dealing honestly with our own ignorance."

— a paraphrased principle, traceable to Karl Popper's philosophy of science

✦✦✦

III. A Mirror Facing Both Ways

One of the central ideas of this book, running through all ten chapters, is that a gaze into the deep past inevitably turns into a mirror pointed back at ourselves.

We searched for hypothetical predecessors — and discovered that the best way to understand what might have remained of them is to ask, honestly, what will remain of us. We searched for alien intelligence in the biology of arthropods and cephalopods — and discovered that our own concept of "intelligence" is dangerously narrow, tailored to the single example we happen to know. We searched for traces of ancient engineering — and discovered that our own science very nearly overlooked large-scale engineering at the very heart of the Amazon, because it didn't match the expected appearance of civilization.

This mirror works both ways, and this is precisely where the book's main practical lesson lies — one that reaches far beyond academic curiosity.

If we allow, even as a cautious, unproven hypothesis, that intelligence and technology may have arisen on Earth more than once, and vanished without a trace more than once, we are compelled to look at our own civilization not as the summit and culmination of the planet's history, but as one more, potentially temporary, episode in a far longer record, far more indifferent to us.

This thought may seem frightening. But it also carries an unexpected liberation: it strips humanity of the illusion of exceptionalism — an illusion that, as we will see further on in this chapter, is sometimes more dangerous than it appears at first glance.

IV. The Ecology of Deep Time

The eighth chapter of this book brought us to an unsettling question: what if the Great Filter — that statistically near-insurmountable barrier separating a technological civilization from its own long-term survival — lies not behind us, deep in evolutionary history, but right in front of us, within the next few centuries?

If so, then the history of Earth we have examined in such detail across nine chapters is not idle archaeological entertainment, but a direct, practical warning. Five mass extinctions. Countless evolutionary lineages that approached the threshold of complexity and never crossed it. A geological eraser, mercilessly wiping away every trace — regardless of whether it was left by triumph or catastrophe.

The planet is built to outlast us regardless. The question is not whether Earth will continue to exist after humanity — of course it will, just as it continued to exist after each of the five mass extinctions. The question is whether we will leave behind only a thin geochemical layer, like the kind we would search for from hypothetical predecessors, or whether we will manage to become that rare exception, the one attempt among many that actually makes it through the filter and continues on, beyond the bounds of a single fragile planet.

In this sense, the entire book is not so much an investigation of the past as a lesson in ecology, delivered on the broadest, longest-term scale imaginable: not the ecology of a single species or a single ecosystem, but the ecology of intelligence itself, set within the history of an entire planet.

"We study the past not because it determines the future, but because it shows us how narrow the path forward really is — and how easy it is to stray from it."

— a closing thought for this chapter

✦✦✦

V. What We Owe Our Descendants — and Strangers We Will Never Meet

The ninth chapter of this book introduced us to the Investigator — a being that may never appear at all, and if it does, will do so three hundred million years from now, when nothing will remain of us but a thin layer of altered chemistry within a body of nameless rock.

There is something deeply moral in the thought that we, right now, with every technological and ecological decision we make, are writing a letter to this hypothetical distant descendant — or, more likely, simply leaving behind a trace that will outlive every memory of us.

This does not mean we are obliged to act "for history" in some grandiose, monumental sense — building pyramids in hopes of eternal glory. As we learned in the sixth chapter, pyramids are not what survives time; what survives time is the modest, inconspicuous geochemical residue of our everyday industrial activity. The grandeur human vanity reaches for almost never coincides with what actually remains.

But it means something more modest, and perhaps more important: an awareness that every collective decision we make — how we treat the planet, how we manage technologies capable of destroying us, how we regard the fragility of our own existence — is inscribed not only into the history of our century or our species, but potentially into the longest history imaginable: the history of Earth itself, a planet that already knows five mass extinctions and perhaps remembers — though we cannot be certain of this — far more than it is willing to show us.

We are, in a sense, indebted not only to our own human descendants, but to that hypothetical, nameless intelligence that might, someday, try to understand us from the scattered fragments of what we leave behind. This is a strange, almost absurd kind of responsibility — a responsibility toward a stranger who may never exist at all. But precisely this responsibility, taken to its ultimate, geological extreme, clarifies better than anything what truly matters.

✦ *A VOICE FROM THE STONE* ✦

We did not write you a letter. But perhaps we should have.

If we had known that someone, after an enormous span of time, would try to understand what we were like, perhaps we would have treated the little we left behind differently. Not tried to leave more of it. But perhaps tried to leave it more honestly.

You still get to choose what your letter will say. We barely had that choice — we were too busy simply living through one more rainy season.

You have a luxury we never had: you know that you are writing a letter. So write it in such a way that you would not be ashamed, if it is ever read.

VI. Strangeness as a Gift, Not a Threat

Let's return to the title of this final chapter. "A Lesson in Humility Before Eternity" — but what exactly does this humility mean in practice, beyond a philosophical pose?

It means a willingness to admit that the history of Earth is deeper and stranger than we're used to thinking — not as a pretty metaphor, but as a literal, geologically confirmed fact. Strangeness here is not a threat to our understanding of the world, but an invitation to expand it.

Throughout this book, we ran into this strangeness again and again: the strangeness of giant dragonflies breathing air unthinkable for modern lungs; the strangeness of the octopus, whose mind grew along an evolutionary path that diverged from ours at the dawn of multicellular life; the strangeness of a forest that was mistaken for wilderness for centuries, though it was a carefully tended garden; the strangeness of a planet that methodically, blindly, with no intention at all, erases its own memory — while, in the rarest cases, preserving tiny, accidental fragments of that memory in the form of concretions, meteorites, and isotopic anomalies.

The strangeness of the world is not an obstacle to science. It is science's very subject matter. Every great discovery in the history of human knowledge began with a willingness to admit that reality is stranger than our habitual intuition suggests — from the roundness of a planet, strange to those who saw nothing around them but a flat plain, to quantum mechanics, strange to those raised in a world governed by Newton's laws.

This book is an invitation to apply that same readiness for strangeness not to the distant physics of elementary particles, but to something far closer and more personal: to the history of our own planet, to the depth of time separating us from the beginning of life, to the possibility that we are not the first, and perhaps not the only ones, ever to have asked ourselves these questions beneath this very sky.

✦ ✦ ✦

VII. One Last Look Into the Dark

Let's picture one final scene — not a scientific one, simply a human one.

Night. You are standing somewhere out beyond the city, far from artificial light, looking up. The sky is strewn with stars — the very sky whose silence we examined in the eighth chapter, the very Fermi Paradox, disheartening and mesmerizing all at once.

But now, armed with everything this book has told you, you can allow yourself to look not only up, but down as well — through the soil beneath your feet, through kilometers of sedimentary rock, down to where lies the testimony of ages that saw this sky long before you did: the Cambrian Explosion, the Carboniferous forests, the Permian catastrophe, the reign of the dinosaurs, their extinction, and the dawn of the mammals.

The universe overhead is silent. But the planet underfoot is not silent at all. It speaks very slowly, very quietly, in the language of isotopes and minerals — a language we are only just beginning to learn to read.

And perhaps the most honest, most beautiful thing that can be said at the end of this book is this: we do not know whether anyone came before us. But we know for certain that the planet we stand on is older than any history we are capable of imagining — and patient enough that it might, one day, tell us the truth, if we are patient enough to listen.

"We look up at the stars in search of company in the universe. Perhaps we should sometimes look down at our feet instead — the company we are searching for may have lived right here, long before we ever learned to look up."

— a closing thought for the book

VIII. A Voice from the Stone, Turned Toward the Future

Across nine chapters, the Voice from the Stone sounded like an echo of the past — an imagined utterance of hypothetical ancient beings, addressed to us, their distant, unwitting heirs. Let this voice, one last time, sound different — not from the past into the present, but from the present into the future. Let this be not their word to us, but our word to those who come after.

✦ *A VOICE FROM THE STONE* ✦

> *If you are reading this — whether a hundred years from now or three hundred million — know this: we knew we were mortal. And we knew we might turn out to be mortal not only as individual people, but as a species, as a civilization, as an experiment of intelligence on this particular planet.*
>
> *We did not know for certain whether anyone came before us. We searched — patiently, sometimes naively, sometimes too cautiously — and found no final answer. But the search itself taught us something important: to look at our own existence not as something to be taken for granted, but as a rare, fragile gift, easily lost and impossible to get back.*

If you have found our trace — whether in the form of cities, of isotopes, or perhaps of this very book, somehow surviving longer than it had any right to — know this: we tried. We made mistakes. We feared disappearing, just as, perhaps, whoever came before us also feared disappearing.

And if you are not us, but someone else, who came in our place because we failed to make it through our own filter — let this letter become, for you, what the fossils of the Carboniferous became for us: not proof, but an invitation. An invitation to look deeper. To ask more honestly. And to remember that the planet's silence is not emptiness, but patience.

✦✦✦

Conclusion

This book began with a pencil, and an instant on its eraser. It ends with that same pencil — only now, we hope, the reader holds it differently: understanding that the thin strip at the tip of the eraser is not merely the whole of human history, but also, perhaps, not the only such strip along the pencil's entire length.

We have not found proof of an ancient civilization. We will probably never find it with complete certainty — such are the laws of geology, the laws of time, the laws of our own limited perception.

But we have found something else, perhaps more valuable: a method of honest questioning in the face of the unknown.

An understanding that absence of evidence is not evidence of absence. Humility before a scale that our minds, which evolved for the far humbler tasks of surviving on the savanna, can barely contain — and, alongside that, a certain pride in the fact that we are nonetheless capable, however haltingly, of reaching toward an understanding of that scale.

The history of Earth is deeper and stranger than we think. Perhaps other voices, long since fallen silent, are hiding behind that strangeness. Perhaps there is nothing behind it at all, save the vast, indifferent silence of stone and time.

Either way, the only sensible response to this kind of uncertainty is neither despair nor overconfident certainty, but attention. Patient, disciplined, grateful attention to the one unambiguous, certain wonder we do possess: that right now, at this very moment, someone — you — is holding this book, breathing, thinking, asking questions beneath the very same sky under which, millions of years ago, someone else may have asked these very same questions.

That, in itself, is no small thing.

THE END

If you enjoyed this book

An eraser doesn't get everything. Sometimes a few words left behind after reading are their own kind of trace — the kind that helps someone else find a book buried under all the others.

If *Boneglyph* gave you something to think about, please consider leaving a short review wherever you bought it. Even a couple of sentences matter — that's usually what other readers are weighing before they decide to open a book at all.

Thank you for reading to the end.

With love and gratitude,

always yours,

David Sterling

Also by David C. Sterling

The Last Child Of Immortality
EL ÚLTIMO HIJO DE LA INMORTALIDAD
L'Ultimo Nato dell'Immortalità
Integration
Интеграция
Integración
O Último Nascido da Imortalidade
Тьма Питается Пустотой
Integração
The Darkness Feeds on the Void
Integrado
La Lasta Infano de Senmorteco
La Oscuridad se Alimenta del Vacío
Starbound Aborigines
Aborygeni dalekich gwiazd
La Mallumo Nutriĝas per la Seneco
Imprint
Indiĝenoj el malproksimaj steloj
Impronta
Afdruk
Impronta
Impronta
İmprint

Entegrasyon
Boneglyph

Watch for more at https://www.amazon.com/stores/author/B0FM67LNTT/about?ingress=0&visitId=2c7fcbb0-1625-4bdb-969b-a667299fb355&ref_=ap_rdr&ccs_id=40d1e6d2-a17b-467f-a604-e2469a363599.

www.ingramcontent.com/pod-product-compliance
Lightning Source LLC
LaVergne TN
LVHW050819061126
851857LV00012B/884